T0135530

FAULT-TOLERANT CONTROL OF
NONDETERMINISTIC INPUT/OUTPUT AUTOMATA

Dissertation zur Erlangung des Grades eines

Doktor-Ingenieurs

der Fakultät für Elektrotechnik und Informationstechnik
an der Ruhr-Universität Bochum

Yannick Salomon Nke
geboren in Mgbaba II, Kamerun
Bochum, 2012

1. Gutachter: Prof. Dr.-Ing. Jan Lunze
             Ruhr-Universität Bochum, Deutschland
2. Gutachter: Prof. Dr. Jean-Jacques Lesage
             École normale supérieure de Cachan, France

Dissertation eingereicht am: 13. November 2012
Tag der mündlichen Prüfung: 01. März 2013

Bibliografische Information der Deutschen Nationalbibliothek

Die Deutsche Nationalbibliothek verzeichnet diese Publikation in der
Deutschen Nationalbibliografie; detaillierte bibliografische Daten sind
im Internet über http://dnb.d-nb.de abrufbar.

ISBN 978-3-8325-3377-9

Logos Verlag Berlin GmbH
Comeniushof, Gubener Str. 47,
10243 Berlin
Tel.: +49 (0)30 42 85 10 90
Fax: +49 (0)30 42 85 10 92
INTERNET: http://www.logos-verlag.de

Acknowledgements

This thesis would not have been possible without the support of several persons that I would like to thank here.

Firstly, I would like to thank my supervisor, Prof. Dr.-Ing. Jan Lunze. Thanks to his experience and critical judgement, I was able to constantly improve my results. He continuously inspired me during technical discussions along the thesis.

I thank Prof. Dr. Jean-Jacques Lesage for accepting to review this thesis. His constructive comments regarding the manuscript as well as during my visit at the Research Laboratory in Cachan have been of great importance.

A special thank goes to the staff of the Institute of Automation and Computer Control at the Ruhr-Universität Bochum, where I have been working during my thesis. This special thank goes to the scientific co-workers Sven Bodenburg, Ozan Demir, Sebastian Drüppel, Jan Falkenhain, Daniel Lehmann, Andrej Mosebach, Jörg Pfahler, Jan Richter, Thorsten Schlage, Melanie Schmidt, René Schuh, Christian Stöcker and Daniel Vey. For the daily technical or administrative tasks I thank Dr.-Ing. Johannes Dastych, Mrs. Kerstin Funke, Mrs. Susanne Malow, Mrs. Andrea Marschall, Dipl.-Ing. Rudolph Pura, Dr.-Ing. Christian Schmid and Dipl.-Ing. Udo Wieser for their support.

I would also like to thank my students who contributed to my works in general, namely, Irfan Garni Altiok, Karsten Bartecki, Sercan Bolat, Jan Eisenburger, Fabian Gerecht, Philip Hohoff, Sven Hügging, Daniel Husslein, Thorsten Kranz, Pablo Ladron de Guevara Lopez, Sanny Mambou, Melanie Schmidt, Daniel Schramm, Philip Weizinger and Kevine Tchaba.

I owe a profound gratitude to my Family and my Friends, who supported me from outside of the university. They have always been there at the right moments. In particular, I have to thank my mother Juliette Ndjana Okala Epse Nke and my father Thomas Nke Ngah for the constant support from the beginning of my education until completing this thesis.

Bochum, March 2013   Yannick Salomon Nke

To my parents.

Contents

Abstract

Problem statement

Given is a faulty plant, a nominal controller and a specification to be fulfilled by the controlled plant. The problem is to find a new controller, for which the closed-loop system satisfies given specifications. This thesis investigates how this problem can be solved if the plant is subject to component faults or failures. The main questions to be answered here are:

1. Under which conditions can the specification be achieved by the faulty system?

2. How should the control law be modified when a fault occurs?

3. How should the controller be implemented to achieve fault tolerance?

These fundamental questions together summarize the fault-tolerant control problem for discrete-event systems, which is the topic of this thesis. In order to answer these questions, methods need to be developed for the following purposes:

- Modeling of the nominal and the faulty plant.

- Control design.

- Analysis of the plant and the control loop.

- Implementation in computer programs.

First, the modeling of the nominal and the faulty plant has to be based on a rigorous formalism. In this thesis, a nondeterministic input/output (I/O) automaton is the discrete-event model selected to describe the system behavior. This model class possesses several advantages compared to other models like standard automata, Petri nets or Markov chains such as the compact and explicit representation of the relation between actuators, sensors and the internal state dynamic of the plant.

However, there are currently no established methods dealing with the control design and fault-tolerant control of nondeterministic I/O automata. For this reason, methods for the design of a controller for the nominal case, the modeling of faults and failures for the faulty case and the controller reconfiguration in a nondeterministic I/O automata framework have to be developed to answer the first question above. The formal elaboration of this framework is a main objective of this thesis. For the second question, the analysis of the control loop requires to study properties like the feasibility of a specification, the controllability and the diagnosability of the plant and the reconfigurability of the control loop. These properties need to be defined in the context of nondeterministic I/O automata. An important analysis step for fault tolerance is the computation of redundancies. However, this thesis shows that the presence of redundancies is not sufficient for fault-tolerant control nor is it always necessary.

Regarding the third question, a computer-aided environment is necessary for modeling, analysis and simulation of discrete-event systems which can easily become very large. In the specific case of fault-tolerant control, it has to be possible to automatically design and reconfigure controllers within a finite time. This will show the applicability of the approach in an industrial environment.

Main results

The method proposed here deals with the systematic reconfiguration of controllers for plants subject to actuator, sensor and system internal faults.

Modeling of faults and failures. This method is developed to derive the behavior of a faulty system out of its nominal behavior and the considered fault in a formal framework. Error relations are introduced for this purpose. This fault modeling technique permits the representation of the behavior of systems subject to several faults and failures affecting the plants behavior simultaneously.

Nominal controller design method. In order to handle a controller reconfiguration in case of a fault, it is necessary to first model the control loop without any fault. This step includes the design of a nominal controller. A realization of this controller and necessary and sufficient conditions for its existence are presented. The latter will be referred to as the controllability condition.

Off-line and on-line reconfiguration method. This is the main result of the thesis. The control design method and the realization structure are extended to a fault-tolerant controller structure. The entities of the controller to be modified are its control policy, its state trajectory and its internal counter that is used to update the target state out of a specified state trajectory. The method is applicable off-line and on-line. The reconfigurability condition derived in this thesis for actuator faults and failures combines the existence of output redundancies in the, so called, supercontroller and the safe feasibility of the specification. For sensor faults and failures, it has been proved that the safe feasibility of the specification in the faulty control loop is necessary and sufficient for the existence of a fault-tolerant controller.

Formal extensions. The *active sets of inputs, outputs or states* introduced in this thesis offer the possibility to precisely select specific signals of an I/O automaton. They are used here to analyze the systems behavior in the nominal and faulty case, to design, to implement and to reconfigure the controller. The control reconfiguration is formalized here by means of *numerical and symbolic adjacency matrices*. These matrices are use here in a way which permits to define a redundancy degree of the controller or determined alternative trajectories from a given state to a target state. In order to visualize those trajectories, *I/O trellis automata* are derived from classic I/O automata through a formal unfolding procedure.

The aforementioned formal extensions are used to express crucial properties as the *safe feasibility* of a specification, the *controllability* of a plant and the *reconfigurability* of a controller. Since the results of this thesis are proven for *nondeterministic I/O automata*, they are also applicable on deterministic I/O automata.

Experimental demonstration and simulation of fault-tolerant control. MATLAB/Simulink programs have been implemented for modeling, controller design, analysis and fault-tolerant control of I/O automata and standard automata. They have been used for simulations and experiments on the three-tank system and the manufacturing cell of the Institute of Automation and Computer Control at the Ruhr-Universität Bochum. Furthermore, the concepts developed in this thesis have been demonstrated on a virtual pick-and-place system and a pilot filling line of the company Siemens AG.

German abstract (Kurzfassung)

Motivation

Die Verlässlichkeit technischer Systeme hängt überwiegend davon ab, wie die Steuerung auf Fehler und Ausfälle bestimmter Komponenten reagiert. Fällt beispielsweise ein Aktor aus, so werden die Stellsignale der Steuerung nicht mehr umgesetzt. Dadurch kommt wird die Strecke entweder zum Stillstand oder weist eine unerwünschte Dynamik im Laufe der Zeit auf. Wenn ein Sensor plötzlich fehlerhafte Werte überträgt, wird die Steuerung ihr Stellsignal solange nicht ändern bzw. falsch reagieren bis der erwartete Messwert antrifft. Solche Fehler können Konsequenzen haben wie Systemzerstörung oder Gefahren für Menschen, falls keine Maßnahmen ergriffen werden. Heuristische Maßnahmen wie Notabschaltungen sind in der Regel mit enormen Kosten, hohem Aufwand und potentiellem Datenverlust verbunden. Deshalb ist es notwendig eine systematische Methodik zu entwickeln, die für eine möglichst breite Systemklasse angewandt werden kann.

Problemstellung

Das Ziel einer fehlertoleranten Steuerung besteht darin, ein fehlerbehaftetes System so zu beeinflussen, dass der Steuerkreis eine vorgegebene Spezifikation erfüllt. Diese Dissertation behandelt den Entwurf solcher Steuerungen und die Analyse von fehlerbehafteten Steuerkreisen. Folgende Fragen werden dabei beantwortet:

1. Unter welchen Bedingungen kann ein fehlerbehaftetes System die Spezifikation weiterhin erfüllen?

2. Wie muss das Reglergesetz angepasst werden?

3. Wie kann der Regler implementiert werden?

Diese Fragen stellen die Kernpunkte dieser Dissertation dar und werden im Folgenden kurz erläutert.

Um die Funktion technischer Systeme aufrecht zu erhalten, sind modellbasierte Methoden zu verwenden aus denen Maßnahmen zur Entgegenwirkung von Fehlern systematisch abgeleitet werden können. Als ereignisdiskrete Modellform werden nichtdeterministische Eingangs-/Ausgangs-Automaten (E/A-Automaten) verwendet. Für diese Modellform existieren keine etablierte Methoden zum Steuerungsentwurf, Fehlermodellierung und Rekonfiguration. Deshalb besteht die erste Aufgabe darin, einen Formalismus für den Steuerungsentwurf, die Fehlermodellierung und die Rekonfiguration von Prozessen, die sich anhand nichtdeterministischer E/A-Automaten beschreiben lassen, aufzustellen.

Eine intuitive Bedingung für die Existenz fehlertoleranter Steuerungen ist generell das Vorhandensein von Redundanzen. Dies muss allerdings mit Hilfe des oben genannten Formalismus nachgewiesen werden. Es gilt zu untersuchen, ob Redundanzen hinreichend und notwendig sind, um die Existenz einer fehlertoleranten Steuerung sicherzustellen. Es ist selbstverständlich, dass redundante Sensoren in der Regel nicht benutzt werden um Aktorfehlern entgegenzuwirken. Deshalb soll die hier entwickelte Rekonfigurationsmethode brauchbare von unbrauchbaren Redundanzen unterscheiden können. Es muss außerdem geklärt werden in welchem Umfang brauchbare Redundanzen zur Erfüllung einer Spezifikation benötigt werden.

Die Erprobung der entwickelten Ansätze kann sowohl durch Simulation als auch durch Experimente erfolgen. Wie es an etablierten Methoden für nichtdeterministische E/A-Automaten fehlt, sind ebenfalls kaum etablierte rechnergestützte Umgebungen zum Entwurf, Analyse und Simulation ereignisdiskreter Steuerkreise, insbesondere nichtdeterministischer E/A-Automaten, zu finden. Ein bekanntes Problem ist dabei die hohe Komplexität ereignisdiskreter Modelle, die oft zur Ausschöpfung der rechnerischen Kapazitäten modernster Rechner führen. Es sind also hierbei keine spezielle Implementierungstechniken zu entwickeln, sondern Konzepte die sich möglichst leicht implementieren lassen. Die praktische Anwendbarkeit der entwickelten Methode werden an realen Anlagen des Lehrstuhls für Automatisierungstechnik und Prozessinformatik an der Ruhr-Universität Bochum demonstriert.

Lösungsansatz

Das in Abb. 1 dargestellte Schema zeigt den strukturellen Aufbau eines fehlertoleranten Regelkreises für ereignisdiskrete Systeme. Die Strecke wird durch einen nichtdeterministischen E/A-Automaten modelliert und befindet sich im Zustand z_p. Tritt das Eingangsereignis v_p ein, reagiert die Strecke mit dem Ausgangsereignis w_p, das von der Steuerung ausgewertet wird um das nächste Stellsignal w_c du generieren. Das Ziel dabei ist die Erfüllung einer Spezifikation durch den Steuerkreis. Der Fehler f wird durch eine konsistenzbasierte Diagnose ermittelt und löst die Rekonfiguration aus.

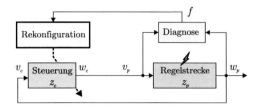

Abbildung 1.: Fehlertoleranter ereignisdiskreter Regelkreis

Die Lösungsschritte des Rekonfigurationsproblems spiegeln sich auf die einzelnen Blöcke von Abb. 1 wider:

1. Modellbildung der Regelstrecke als nichtdeterministischer E/A-Automat und Entwicklung von Funktionen mit denen die in Abb. 1 vorkommenden Signale einzeln erfasst werden können.

2. Entwicklung einer Fehlermodellierungsmethode für Aktorfehler, Sensorfehler, interne Systemfehler und Ausfälle.

3. Entwicklung einer Steuerungsentwurfsmethode. Dafür sind mögliche Spezifikationen, die an den Regelkreis gestellt werden können, zu definieren und zu modellieren.

4. Herleitung einer Rekonfigurationsmethode ereignisdiskreter Steuerungen, die sich sowohl off-line als auch on-line anwenden lässt.

Im Folgenden, werden die für jeden Schritt erzielten Ergebnisse erläutert.

Hauptergebnisse

Formalismus zum Entwurf fehlertoleranter Steuerungen anhand von E/A-Automaten.
Der hier entwickelte Formalismus beruht auf der Boolschen Algebra, der Mengentheorie und der ereignisdiskreten Systemtheorie. Dafür werden hauptsächlich charakteristische

Funktionen, aktive Mengen, boolsche Operatoren, numerische und symbolische Adjazenzmatrizen eingesetzt. Der aktive Nachfolgezustandsoperator $\mathcal{Z}_{ap}(\cdot)$ wird z.B. in Abb. 2 benutzt um die mögliche Nachfolgezustände des Automaten $\mathcal{N}_p$ für eine gegebene Zustandsmenge $\mathcal{Z}_k$ und die aktuelle Eingabe v_k zu bestimmen. Mit Hilfe von symbolischen Adjazenzmatrizen lassen sich Zustandsfolgen zwischen zwei Zuständen ablesen. Eine Kombination aus aktiven Mengen, boolschen Operatoren und charakteristischen Funktionen wird beispielsweise benutzt um Redundanzgrade zu berechnen.

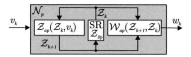

Abbildung 2.: Struktur des fehlerfreien E/A-Automaten

Fehlermodellierungsansatz. Mit dem durch Abb. 3 beschriebenen Ansatz wird gezeigt, wie das Verhalten eines fehlerbehafteten Systems modelliert werden kann. Mithilfe des Modells des fehlerfreien Systems und der eingeführten Fehlereingangsfunktion $E_v(\cdot)$, der Fehlerausgangsfunktion $E_w(\cdot)$ und der Fehlzustandsfunktion $E_{z'}(\cdot)$ wird die charakteristische Funktion der fehlerhaften Strecke berechnet. Der Ansatz ermöglicht die systematische Modellierung von Aktor-, Sensor- und systeminternen Fehlern und Ausfällen, die gleichzeitig auf das System wirken.

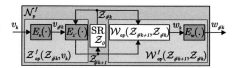

Abbildung 3.: Struktur des E/A-Automaten der fehlerbehafteten Strecke

Steuerungsentwurfsmethode. Im fehlerfreien Fall wird die Steuerung aus dem Modell der Strecke $\mathcal{N}_p$ und der Spezifikation $\mathcal{S}$, wie in Abb. 4, abgeleitet. Zur Überprüfung der Erfüllbarkeit einer Spezifikation wird ein Kriterium entwickelt, mit dem die absolute Garantie der Erfüllbarkeit einer Spezifikation überprüft werden kann. Ein wichtiger Schritt dabei ist die Bildung des Spezifikationsautomaten $\mathcal{N}_s$, durch den in Abb. 4 mit Spec$(\cdot)$ bezeichneten Schritt. $\mathcal{N}_s$ stellt den reduzierten Automaten dar, der sich ergibt, wenn die laut der Spezifikation verbotenen Transitionen und Zustände aus dem Modell der Strecke entfernt werden.

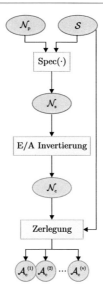

Abbildung 4.: Steuerungsentwurfsablauf

Zur Realisierung der berechneten Steuerung wird eine Struktur erarbeitet, mit der der nominelle und der fehlerhafte Steuerkreis untersucht werden kann. Diese Struktur ist im unteren Blockschaltbild von Abb. 5 dargestellt. Sie besteht aus

- einem Vergleichszähler k_s, der nur dann inkrementiert wird, wenn der aktuelle Messwert $w_p(k)$ mit dem erwarteten $\hat{v}_c(k)$ übereinstimmt,

- der Solltrajektorie $\overline{Z}_s(0 \cdots k_e)$, die von der Strecke verfolgt werden soll und

- dem Steuergesetz $\mathcal{A}_c^{(i)}$, das die für eine Transition notwendigen Stellsignale und erwartete Messwerte beinhaltet.

Steuerungsrekonfigurationsmethode. Die Realisierungsarchitektur der Steuerung ist durch eine Schnittstelle erweitert worden, die die Rekonfigurationsbefehle in spezifische Signale umformt, die von der Steuerung umgesetzt werden können. Dabei geht es um eine Anpassung des Steuergesetzes, der Zustandstrajektorie und des internen Zählers der Steuerung, mit dem die Steuerung mit der Strecke synchronisiert wird. Diese Methode lässt sich sowohl off-line als auch on-line anwenden. Die Rekonfigurierbarkeitsbedingung ist getrennt für Aktorfehler und Sensorfehler aufgestellt worden. Bei Aktorfehlern und Aktorausfällen ist die Existenz von Redundanzen und die Erfüllbarkeit der Spezifikation notwendig und hinreichend. Bei Sensorfehlern und Sensorausfällen konnte gezeigt werden, dass die Erfüllbarkeit der Spezifikation notwendig und hinreichend für die Existenz einer fehlertoleranten Steuerung ist.

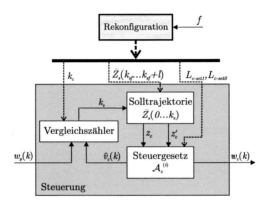

Abbildung 5.: Rekonfiguration der Steuerungselemente

Experimentelle und simulative Erprobung. Im Rahmen dieser Dissertation sind zur Simulation und Erprobung der obigen Konzepte Programme in MATLAB/Simulink implementiert worden. Sie ermöglichen die Modellierung, den Steuerungsentwurf, die Analyse, die fehlertolerante Steuerung und die Visualisierung von nichtdeterministischen E/A-Automaten sowie Standardautomaten. Sie sind bei experimentellen Versuchen am Dreitanksystem (Abb. 6) und der Fertigungsanlage (Abb. 7) des Lehrstuhls für Automatisierungstechnik und Prozessinformatik an der Ruhr-Universität Bochum eingesetzt worden.

Im Folgenden werden einige Ergebnisse der Rekonfigurationsmethode an der Fertigungsanlage erläutert. Der nominelle Fertigungsprozess besteht darin, die Werkstücke vom Platz 15 zum Platz 7 der Anlage zu transportieren. Abbildung 8 zeigt die Aufnahmen der Stellsignale V_p, der Messsignale W_p sowie die interne Zustände $\hat{Z}_p$ der Anlage.

Spezifiziert wurden sechs Zustände von $Z_s(2)$ bis $Z_s(12)$. Zwischenzustandsfolgen wie z.B. von $Z_s(2)$ bis $Z_s(4)$ stellen mögliche Zustandstrajektorie zwischen den spezifizierten Zuständen dar und werden als Freiheitsgrade der Spezifikation aufgefasst. Dementsprechend hängen diese Zustandsfolgen und die dazugehörigen Ein- und Ausgabefolgen vom

Abbildung 6.: Foto des Dreitanksystems des Lehrstuhls für Automatisierungstechnik und Prozessinformatik an der Ruhr-Universität Bochum

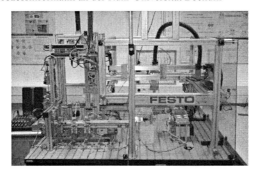

Abbildung 7.: Foto der Fertigungsanlage des Lehrstuhls für Automatisierungstechnik und Prozessinformatik an der Ruhr-Universität Bochum

aktuellen Steuergesetz ab. Abbildung 8 zeigt also das Verhalten des Regelkreises mit einer Steuerung, die aus mehreren ausgewählt wurde.

In Abb. 9 wird das Verhalten des fehlerbehafteten Regelkreises vor und nach der Rekonfiguration dargestellt. Zum Zeitpunkt $t_f = 7\,s$ wird ein interner Systemfehler diagnostiziert. Der Platz 7 ist nämlich durch ein anderes Werkstück belegt. Die Rekonfiguration wird angestoßen (siehe Abb. 9). Die neue Steuerung ist bei $t_r = 93\,s$ berechnet und führt Rekonfigurationsschritte durch bis der nominelle Prozess ab $t_n = 89.5s$ fortgesetzt wird.

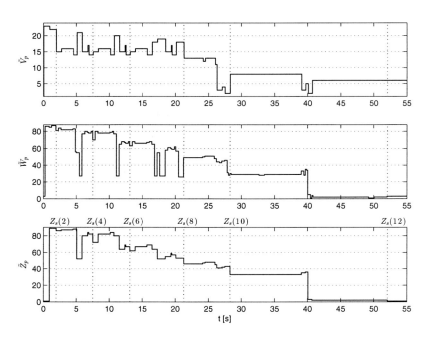

Abbildung 8.: Nomineller Transport eines Werkstücks vom Platz 15 zum Platz 7 der Fertigungsanlage

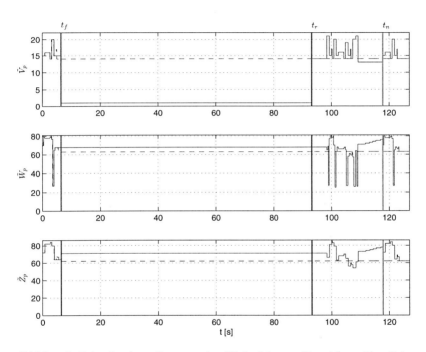

Abbildung 9.: Rekonfigurierter Transport eines Werkstücks vom Platz 15 zum Platz 7 der Fertigungsanlage

1. Introduction

Abstract. *This chapter gives a motivation of fault-tolerant control from a theoretical and practical view point. Fundamental issues ranging from error detection to fault treatment of discrete-event systems are described to precisely formulate the fault-tolerant control problem of this thesis. A literature overview presents the state of the art in the field of fault-tolerant control of discrete-event systems. The solutions developed in this thesis are summarized. A manufacturing system and a fluid level control process which serve as running examples to illustrate the methods are introduced.*

Dynamical systems consist of actuators, sensors and other physical components to fulfill a designed task. In the nominal case, which is characterized by the absence of faults, the controller enforces the execution of this task by sending appropriate commands to the actuators. The effects of these commands are captured by the sensors and fed back to the controller which generates the next command according to the given task.

Faults or failures may change this interaction in several ways. They can either occur at the actuator level, the sensor level or in the other components of the system. A broken cable may disconnect a plant from its controller such that some commands and measurements are no longer communicated among each other. Electromagnetic interferences may provoke a signal falsification or delay in the communication between the plant and its controller. As a result, the plant reacts in an unexpected way and the behavior of the control loop becomes unpredictable. This malfunction can end up into damages of the system and its environment. Hence, there is a need to develop a fault-tolerant control scheme which inhibits the influence of a fault and permits the achievement of the designed task.

This thesis deals with the reconfiguration problem of discrete-event systems shown in Fig. 1.1. Input and output signals are used to describe the interaction of the system with the controller. The fault is assumed to alter the input/output behavior (I/O behavior) of the system. The measurements of the input and output of the system are compared in the diagnostic unit with a model of the system's behavior.

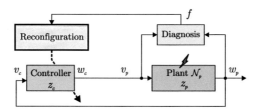

Figure 1.1.: Fault-tolerant control framework for I/O automata

1.1. Discrete-event control reconfiguration problem

The fault-tolerant control framework considered in this thesis is illustrated in Fig. 1.1. It consists of a discrete-event system (DES) modeled by an I/O automaton $\mathcal{N}_p$ which is in the state z_p and has to execute a given task. The description of this task is formulated by a specification to be fulfilled by the closed-loop. The plant $\mathcal{N}_p$ receives commands v_p in form of input events and reacts with output events w_p. The latter is used by the controller to produce the next control event $w_c = v_p$ depending on the specification to achieve. After a fault, the control loop may block because of unexpected events being communicated. Both input and output signals of the plant are processed by the diagnoser in order to detect and identify the fault f. The fault diagnosis triggers the reconfiguration which modifies the control law in a way that the faulty closed-loop still fulfills the specification. The objective of reconfiguration can be broken down into the following tasks:

1. How to find a new trajectory for the faulty plant in order to achieve the specification?

2. How to derive a controller from the new trajectory that guarantees nonblockingness of the control loop?

The solution of the first task depends on the model of the faulty plant and the specification to be fulfilled. For instance, if the faulty plant is modeled by an automaton and the specification is a target state to reach, the task reduces to a well-known graph search problem. However, this thesis does not focus on the graph search method but on the formalization of discrete-event control reconfiguration method based on nondeterministic I/O automata.

The second task is based on the assumption that a trajectory exists e.g. from a faulty state to a target state. However, there is no guarantee that the new trajectory is completely safe until the target state. The solution of this safety issue relies in the conditions under which the faulty plant can execute the new trajectory without any deviation that would prevent it from reaching the target state. These conditions describe the reconfigurability of the control loop.

If the reconfigurability condition is satisfied, the methodology should suggest control reconfiguration steps from which a new controller can be derived. The correcting actions undertaken by the new controller can lead to full recovery, a degraded recovery or a safe shutdown according to the system and the specification at hand [36]. A failed reconfiguration leads to a system failure represented by a deadlock, a livelock or a disintegration.

1.2. Motivation

1.2.1. Missing methods for fault-tolerant control with I/O automata

The fault-tolerant control problem has been extensively addressed for monolithic continuous models as in [128]. The recently published survey of [65] discusses new results and challenges in the field of fault-tolerant control of continuous systems under network considerations. However, the interaction of most systems is nowadays digitalized in bits sequences. An appropriate modeling framework for digital systems is the discrete-event modeling framework which includes standard automata, input/output automata, states machines, Petri nets and other similar models.

Compared to standard automata, I/O automata explicitly better model the causality property which is fundamental in technological systems. The behavior of many real-world systems or phenomena are nondeterministic in nature. Especially, the occurrence and behavior of faults can not be predicted in advance. On the one side, solutions for nondeterministic problems and analysis methods are usually more complex than the deterministic ones. On the other hand, every result that holds for nondeterministic I/O automata also holds for deterministic I/O automata. The contrary is not always possible.

Therefore, this thesis is devoted to input/output (I/O) automata which have not yet been used for fault-tolerant control. In [77], an emphasis is put on the nature of the model of the component involved in the automation process, i.e, if they are continuous, hybrid or discrete. In the context of this thesis, all components are purely discrete. Since the use of I/O automata for fault-tolerant control is new, there are several issues emerging while trying to setup formal, systematic and general control reconfiguration solutions:

- How should verbal requirements of a task to execute be formalized as a *model specification*? Under which conditions can this specification be fulfilled?

- How should degrees of freedom implicitly and explicitly be involved in the task definition? Are they always necessary to achieve fault tolerance?

- How to derive a discrete-event controller of a nonfaulty system for a specific task in the I/O automata framework?

- What are the consequences on the plant and how should they be modeled once a fault has been diagnosed?

- How should the reconfiguration of the discrete-event controller be performed?

This thesis proposes answers to the questions stated above in a formalized discrete-event system framework. These questions particularly concern the steps needed to achieve a control reconfiguration in an I/O automata framework. Since the realization of each step needs some efforts, it is worthwhile to investigate under which conditions they can be achieved successfully. These conditions need to be formalized as system properties:

- **Feasibility of a task by a controlled system.** Before a possibly intensive computation of the controller, it is important to test if the required specification can be achieved by the system in a closed-loop with the controller. In the positive case, the specification is said to be *feasible* in the system otherwise it is *unfeasible*.

- **Controllability of a plant for a given specification.** The feasibility of a specification by a system does not guarantee a deterministic control of the system by a controller. The existence of a controller able to deterministically enforce the achievement of the specification by the plant represents the controllability of the considered nonfaulty plant. Note that feasibility of the specification is only a necessary condition. A sufficient condition is formally presented in this thesis.

- **Reconfigurability of a controller for a given fault.** The reconfigurability as a system property is explicitly addressed in [76] for hybrid systems. The systems are hybrid in the sense that their dynamic is continuous but the faults are discrete events. At the actual state of research, the reconfigurability property is still not formalized for I/O automata. In this thesis, it is formalized as the controllability of faulty systems for the same specification as in the nominal case.

The main theoretical motivation can be summarized in the need of formalized steps to achieve fault-tolerant control in I/O automata but also the conditions under which these steps can be performed successfully.

1.2.2. Practical motivation

Heuristic checkpoints declaration. In manufacturing processes Programmable Logic Controllers (PLCs) usually implement step chains in order to realize a required product. In the case of a fault, e.g., memory overflow, cable break etc, the CPU of the PLC may go into a STOP mode with a potential loss of data. A severe economic loss could be a consequence. This situation is usually technically solved by the heuristic declaration of checkpoint states based on an expertise for the considered process. The first limitation of this approach is encountered when system or process modifications are performed because new check points have to be explored again. This problem becomes untractable for large scale plants. Thus, a systematic reusable methodology helps to reduce the time consuming search of checkpoints after every system or process modification. The method would just induce an adaptation of the control law to fault. However, a minimal number of checkpoint states is sometimes required during system design by safety constraints. Hence, a minimal number of checkpoints need to be taken into account by the methodology. *How should the control law of a faulty system be modified without use of checkpoints, so that e.g. a stopped manufacturing process can be resumed?*

Challenges of autonomy. Consider environments which are hardly accessible to humans, like deep undersea or space oriented systems like satellites, interplanetary exploring robots, etc. If a component is malfunctioning or broken, it is expensive and time consuming to setup a team of experts to fix the problem. Instead, there should be an autonomous control reconfiguration procedure which is launched once a fault is diagnosed and takes over the control of the faulty plant until a team of experts intervene. A solution could be the development of a robust controller but its applicability is diminished by limited memory and computation capacities in those environments. A better solution would be to reconfigure the controller on-line solely for the fault at hand instead of trying to implement a controller for any fault in advance. *How should the reconfiguration of a controller be undertaken on-line whilst taking care of limited memory and computation capacities?*

The questions raised in this section reflect important problems encountered when trying to implement theoretical methods in real-world applications. Thus, it motivates the search for answers to these questions in the following.

1.2.3. Fundamental issues of fault-tolerant control

In this thesis, the fundamental mode of operation is assumed in the sense of [107] because only one variable of state, input or output is allowed to change at any time. In this thesis, a fault-tolerant control system is an autonomous control loop extended with a diagnosis and control reconfiguration unit which is able to adapt the control law so that the plant still achieve a predefined task in the presence of a given set of faults and failures. Reference [119] proposes to divide the fault-tolerant control problem in four key principles as follows:

1. **Error detection:** This will be referred to fault diagnosis and identification (FDI) which pursue the objective of detecting, localizing and identifying faults and failures of the plant within a finite number of steps. This is not in focus of this thesis, hence, it is assumed to be handled by the diagnoser in Fig. 1.1.

2. **Damage confinement and assessment:** This step consists of checking how far the system may have been damaged since the FDI result is known and to which extent further damages may occur with respect to the dynamic and structural properties of the plant. This issue is tackled by the fault modeling approach developed in Chapter 3.

3. **Error recovery:** The main goal is to drive the system from a faulty state back to a functioning state from which the specification can be achieved as in the nominal case. This is the core of this thesis. It is applicable in off-line and on-line operation of the control loop.

4. **Fault treatment and continued service:** This concerns measures intending to exclude recidivist effects of faults in the post-reconfiguration period of the control loop. This technique will be presented for off-line reconfiguration only, because of the highly demanding resources which are assumed to be unavailable on-line.

Crucial system properties for this thesis are now introduced from [37] and [119].

Dependability. It is the property of a system which allows reliance to be justifiably placed the service it delivers.

Reliability. It is the ability of a system to deliver its normal service. In this thesis, the normal service will be modeled by a specification.

Availability. It is the percentage of time for which the system is delivering its service.

The objective of this thesis is to propose methods of fault-tolerant control in order to improve the reliability of technical processes modeled by discrete-event systems in the presence of faults and failures. The methods presented here could result in an improvement of the availability of the considered systems even though it is beyond the topic of this thesis.

The reliability problem of technological systems has gained interest since the early development of the computers, titled at that time as "calculating machines" [26, 155, 188] who pointed that issue out and prognosed it to become relevant in the future.

1.3. Application examples

The application of concepts of the fault-tolerant control are various. Some academic examples of fault-tolerant control of discrete-event systems are given in this section. Section 1.3.1 describes only those processes where the methods presented in this thesis can be studied manually. The manufacturing cell of the Institute of Automation and Computer Control at the Ruhr-Universität Bochum is also used in this thesis as an application example. It is not presented in the following sections but in Chapter 7 due to its complexity. Section 1.3.2 presents other examples from the literature where the methods of this thesis can also be applied.

1.3.1. Examples used in this thesis

Level control of a fluid process

Consider the plant performing a fluid level control process shown in Fig. 1.2.

Nominal level control. The control objective of the process is to fill the tank T_1 from level 0 up to level 4, then down to level 1 and back to level 4 in a cyclic way.

The fault-tolerant control objective is to achieve the previously described cyclic process despite the fault "valve V_2 stuck at closed". The solution is trivial. The valve V_3 must be used when valve V_2 blocks and the control law has to be adapted accordingly. The simplicity of this example is used to demonstrate the applicability of a general formal solution in a specific case. It shows how the solution to the reconfiguration problem can be found automatically by the method developed here.

Faulty level control. When the valve V_2 is malfunctioning, the consequences depend on the dynamics of the plant, the behavior of the controller and the nature of the fault. V_2 influences the dynamics of the plant in two ways. It reduces the level of educt in the

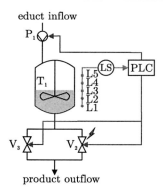

educt inflow

product outflow

Figure 1.2.: Mixture preparation process

tank T_1 when opened or keeps the level of educt constant when closed. A malfunction of V_2 could result into a modification of this influence on the process. The behavior of the controller plays an important role into the detection of the fault because it can amplify or mask its effects. A controller which only makes use of valve V_3 would mask the effects of the fault "valve V_2 stuck at closed" but not those of the fault "valve V_2 stuck at open". This shows the importance of the relation between the nature of the fault, the current actions of the controller and the own dynamics of the plant. It also shows that a heuristic analysis of the behavior of the control loop demands costly efforts compared to the systematic fault modeling approach explained in Chapter 3. In the example of this thesis, the fault "valve V_2 stuck at closed" leads to a stagnation of educt at level $L4$ and a blocking nominal controller which continuously tries to open the valve V_2.

Reconfigured level control. Once the fault "valve V_2 stuck at closed" has been diagnosed, a model of the faulty plant $\mathcal{N}_p^f$ is constructed to derive a new control law based on the specification. The methodology is presented later in an off-line and on-line realization. However, the expected result is obvious. In the control law, the command "activate actuator 2 and deactivate the others" is replaced by the command "activate actuator 3 and deactivate the others". The off-line control reconfiguration replaces *every transition* where the activation of V_2 is foreseen by the activation of V_3 in *one step*. The on-line control reconfiguration replaces the activation of V_2 by the activation of V_3 only for *the blocking transition*, i.e, every time the control loop is blocking. Hence, the on-line reconfiguration requires *many steps* which finally lead to the same controller obtained through the off-line reconfiguration.

Manufacturing process: Pick & Place system (PPS)

The Pick & Place system (PPS) presented next is a simplified version of the one presented in [161]. The system consists of two conveyor belts and a magnetic gripper depicted in Fig. 1.3. The conveyor belt B_1 supplies the system with three kind of workpieces: triangular, round and square with the IDs 1, 2 and 3, respectively. The workpieces have to be placed in empty holes of the box on the conveyor belt B_2. The transport of the workpieces in the PPS is performed by the magnetic gripper. A visual sensor *VS* placed at the left-hand side of B_1 permits to detect the ID number of the current workpiece. The gripper is equipped with a thermal infrared sensor TS enabling the measurement of the temperature of the workpieces.

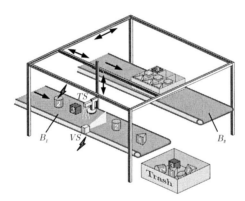

Figure 1.3.: Pick & Place system

Nominal manufacturing process. The process consists of picking the workpieces from B_1 and place them in the box on B_2. It is hereby necessary to sort the workpieces in the shown order $1 - 1 - 1 - 2 - 2 - 2 - 3 - 3 - 3$ in a cyclic way. If the detected workpiece on B_1 is not the expected one, it is not transported by the gripper to the box and, hence, falls into the trash. Assume that the workpieces 1, 2 and 3 always enter and stay in the system with a temperature of $10°C, 20°C$ and $30°C$ respectively. The surrounding temperature is assumed to be $37°C$.

Faulty manufacturing process. Two faulty scenarios are considered and labeled as the fault f_1 and f_2.

- f_1 is an internal system fault where the supply order of workpieces suddenly changes randomly. That is, the expected order of workpieces on B_1 is no longer respected. As a consequence, the nominal controller would wait for the "right" workpiece to come and let the others fall into the trash meanwhile. With respect to the number of "wrong" workpieces falling into the trash, the energy consumption and the eventually long dead times, the production efficiency of the PPS would decrease.

- f_2 is an aggravation of f_1 through a dysfunction of the visual sensor VS which is stuck at the value 2 once the first column of the box is full. Thus, the nominal controller would consider any workpiece as the one with the ID number 2. This would result in the situation where the second column of the box on B_2 is full with eventually wrong pieces whereas the third column remains empty. The specification will be violated.

Reconfigured manufacturing process. There are redundancies in the system that can be used to hold the faulty plant in operation. The reconfiguration method developed in the next section should find, for example, the following reconfiguration solution:

- After the internal system fault f_1, the reconfigured controller should either transport the wrong workpieces to the trash or move them to the right column of the box.

- A solution to the visual sensor fault f_2 would be to use the thermal infrared sensor of the gripper TS to deduce the ID number of the workpieces depending on their temperature. The reconfigured controller should then ignore the measurement of the visual sensor but use those of the thermal sensor to keep the process running.

The aim of this example is to demonstrate the applicability of the fault-tolerant control method developed here in a more complex example than the previous one. Indeed, the resulting model of the PPS consists of 64 states and 209 transitions compared to 11 states and 59 transitions for the level control process. This reflects the necessary efforts needed for a heuristic approach compared to the method developed here. This complexity issue is emphasized on the fault-tolerant control of a real plant manufacturing cell at the Institute of Automation and Computer Control at the Ruhr-Universität Bochum. The method was extensively applied in [20, 25] in simulations and experiments where the storage and computational capacities of modern computers have been exhausted. This was not due to control reconfiguration method applied but to the well-known state space explosion problem of discrete-event systems.

1.3.2. Examples from the literature

Self-reconfigurable robots. These robots are modularly constituted of components which enable the controller a change of their shape or morphology depending on the task to achieve. For instance, such a robot can change from a crane to a car and transport work-pieces. Each module has its own actuators, sensors, batteries and processor, so that it can disconnect, move and reconnect via eventually active connectors. The reconfiguration steps mostly consist of automatically disconnecting, moving and reconnecting modules so that the new shape of the robot permits to perform a task [174, 183, 184]. It is more efficient to define the functionalities of each shape in advance in order to compute the suitable ones during the reconfiguration. Most of the self-reconfigurable robots are controlled by a central unit which decides which shape the robot should take to perform a given task.

Conventional controllers which are designed for specific tasks are usually optimized and straight forward realized. It is not the aim of the reconfigured controller presented here to compete with those but to propose an alternative controller design method which represents a help to achieve a specification when the original optimal controller become useless due to the fault.

Smart grids. They increase the reliability of power facilities by combining remote fault-tolerant control with energy production, transmission and distribution with low outage rates [47, 69, 135]. In addition, the complexity of smart grids can be managed with I/O automata which are the models used in this thesis. The polynomial complexity of the reconfiguration algorithms presented here enables a day ahead as well as real-time re-scheduling of energy production according to the faults.

Smart grids are a subdomain of smart structures which have the ability to respond to changes internal and external conditions. Reference [134] underlines the analogy between smart structures and biological systems as the former tend to satisfy several characteristics of the later as sensing, actuation, adaptability and self-repair.

1.4. Literature overview

1.4.1. Fault tolerance paradigms

Design of fault-tolerant systems. A design method of fault-tolerant systems is proposed in [80] by means of degraded specifications, synchronous observers to cope with sensor faults and new states addition to augment redundancies. An overview of related methods in the field of design of fault-tolerant systems is also given therein. References [33, 34, 51, 68, 78, 108] and [109] focus on the design of fault-tolerant systems by means of detectors and correctors. Detectors are components preserving a safety condition, i.e, the system performs only safe transitions. In case of a fault, the invariant of detectors is false and the corresponding action of the system can not be executed. Correctors are detectors extended with the ability to ensure the liveness condition, i.e, they are able to drive the system into a state from which the specification can be fulfilled, e.g., reach a target state. The major differences and similarities with the approach of this thesis are the following aspects:

- Differences

  - Approach and complexity: This thesis is devoted to the on-line reconfiguration of a controller without any explicitly pre-defined reconfigurations steps as it is the case for the design of fault-tolerant systems. Note that the pre-definition of detectors and correctors for each action and each fault is typical for passive fault tolerance whereas this thesis presents active fault tolerance. In addition, exponential complexity issues emerge when detectors and correctors have to be added at the atomicity level of the model considered here or for cyclic processes. Instead, the reconfiguration algorithm presented here has a polynomial complexity.

  - The interpretation of safety: The safety condition as defined in [33] and [68] excludes sequences with bad prefixes or bad transitions, whereas they are acceptable in this thesis under certain conditions, namely, if a violation of a specification by a plant can be deterministically avoided by a controller. That is, for any prefix, bad suffixes can be excluded from the plants dynamic through adequate inputs.

  - The level of fault tolerance: In the design of fault-tolerant systems, the designer has the possibility to set a level of tolerance for a set of fault hypotheses. [68, 80] and [109] propose efficient methods to add redundancies efficiently.

Instead, this thesis proposes an approach to deal with the redundancies at hand without adding some more.

- Similarities

  - The consideration of traces or sequences of states or events to express important properties like safety and liveness.

  - The definition of fault tolerance: [33] and [68] define it as the ability of the plant to remain in a bad state due to fault, but from which the good states are still reachable. In this thesis, the fault tolerance is defined for a control loop. However, both definitions are similar in the sense that bad states from which good states are not reachable violate the fault tolerance in both cases.

The design and development issues of systems which are fault-tolerant by construction is also investigated from the view point of engineering with less emphasis on formal methods but an extensive demonstration on two benchmark processes: a diesel engine actuator [48] and the Ørsted Satellite [40].

Topological fault tolerance. Reference [36] explains the idea of fault tolerance from the physical, architectural and implementation point of view. It states that "Fault tolerance is the unique attribute of a digital system which makes it possible for the system to continue with its program-specified behavior as a logic machine after the occurrence of faults." However only topological issues of fault tolerance are discussed in [36]. The belief that redundancies are necessary for fault tolerance is particularly emphasized or considered as granted. This thesis shows how functional redundancies can be used where physical redundancies (in terms of different components achieving the same task) are not available.

Mechanical and structural fault tolerance. Faults and failures in mechanical components are usually seen as damages because some material may lose its initial property, be degraded or destroyed [100, 143, 168]. Common manifestations are cracking or fatigue of components. Solutions intended to deal with structural damages are mainly inspired from the structural health monitoring (SHM) community [41]. A new paradigm is to consider the current fault, the actual load and the environment of the system to forecast future behaviors and anticipate with correcting actions. Damage prognosis is used in [73] and [100] to investigate this paradigm of fault tolerance. A bridge between the former and methods of fault-tolerant control is a new field called damage fault tolerant active control [134].

Fault-tolerant supervisory control. The fault-tolerant control of standard automata under sensor failures is discussed in [165] whereas a formal framework is proposed in [110]. The existence and synthesis problem of a controller for the plant with broken sensors is reformulated as an observability test as in [123]. Further approaches of this field are classified and presented in Section 1.4.2 since the methodology is closer to this thesis.

Biological fault tolerance. Discrete-event systems are suitable not only for technological systems but also for biological [115, 175, 181], economic, sociological and psychological investigations. In biological systems, tolerance is defined as the unresponsiveness of an element of the immune system to an antigen (non-self cell). It results from a selection of elements of the immune system which can interact with antigens (structures of the pathogens) without being responsive to autoantigens (structures of the host or self cell) [50, 74, 189]. [189] summarizes analogies between immunological systems and fault-tolerant programs. An implementation of an immune system on a FPGA is proposed in [50].

Cognitive fault tolerance. The cognitive impact of modifying the word sequences in terms of understanding, speed of recognition and difficulty of recognition is studied and discussed in [84, 158, 159] and [160]. The technical reformulation of this situation is quite similar. While investigating the fault-tolerant property of a system, the impact of faults, the speed of recovery and the quality or degradation of its performance after the reconfiguration are also of interest. Consider for example the sequences:

fault-tolerant   control

flaut-tloreant   cnotrol

faut-toleramt   comtrol

They have different patterns but certain similarities such as ordering, phonetics and also different letters. However, since the brain of the reader is expecting the correct sequence "fault-tolerant control", the sense of faulty words may still be understandable. Note that the faulty words either have a wrong order or a different letters, but the initial and final letters are always remain unchanged as mentioned in [158, 159], and [160]. The faults and failures considered in this thesis will obey the same logic. Faulty signals are assumed to result from a signal falsification via appropriate functions.

Further fault-tolerant control approaches are given in [36] and [153].

1.4.2. Fault-tolerant control of discrete-event systems

Tolerance automata. One of the first modeling approaches dealing with fault tolerance of DES was developed by [32] with tolerance I/O automata. This concept has been used later by [61] to propose a framework based on finite-state machines to analyze typical fault-tolerant behaviors such as masking, fail-stop, t-fail-stop, and degradable fault tolerance.

Optimal control. A solution of the optimality problem in tolerance automata has been proposed in [185] by means of a tolerance cost automaton. [154] defines the fault tolerance as a false-alarm-minimizing strategy and proposed a method to cope with sensor faults. [187] addressed the optimal fault-tolerant control problem from another perspective, where the set of nonfaulty behaviors is maximized and the set of faulty behaviors prior the recovery is minimized. [67] used methods of optimal discrete controller synthesis to build a system which is fault-tolerant and guarantees real-time constraints by construction. [98] and [99] focuses on the optimal selection of a single event to enable while minimizing the execution costs to reach a target state.

Switching and robust control. The switching control paradigm has been followed for standard automata by [144] with the concepts of safe diagnosability and safe controllability as necessary conditions in order to achieve a fault-tolerant supervision by investigating the so-called bad states once a safe diagnoser has been built. In [110] the switching from the nominal controller to the fault-tolerant controller is implicitly achieved by the fault event, i.e, at the fault detection step. The fault event triggers a switch to a new or usually degraded specification for which the fault-tolerant controller is designed. A combination of fault-tolerant robust supervisory control of DES with time issues is also discussed in [147], [146] with the concept of Tolerable Fault Event Sequence (TFES).

Petri nets and implementation. A state avoidance approach with observer-based methods of Petri nets is presented in [190] for fault tolerance issues. Implementation techniques for reconfiguring a distributed information fusion system are presented in [45] based on a runtime model. The model of the runtime system is based on a Generalized Stochastic Petri Net (GPSPN) and the computation uses the resulting Markov chain.

Corrective control of sequential machines. The fault-tolerant control problem is presented in [89] as a corrective control problem where the objective is to design an open-loop controller that drives the plant from an unknown initial condition to a target steady-state. This concept has its origin in [87, 88] and is further developed in [79, 90, 91, 136, 137, 148,

179, 180, 192] and [193]. The corrective control problem considers a plant which operates without a controller until a fault occurs due to inadequate inputs from the environment. The corrective controller is then used as an autonomous component which ignores the current state of the faulty plant. The authors use a nonlinear framework in an input/output sequential reasoning. The reconfigurability condition is explained as the possibility to correct the behavior of the faulty plant before it is too late, i.e., before the system reaches failure state representing a point of no return. No model of the faulty behavior is investigated but the initial state of the faulty plant is assumed to be unknown.

Degree of fault tolerance. An important question for the reconfigurability is an estimation of the degree of fault tolerance. That is to which extent certain faults are tolerable, i.e, do not jeopardize the achievement of the specification, so that a reconfiguration is still possible. The problem is to find out which severity of the fault is acceptable. [173] studies this question by proposing an approach to measure the usefulness of an actuator or a sensor when a fault occurs. In a top-down representation of available components in the system, the authors propose to count the number of actuators or sensors available to maintain properties like the controllability and the observability, respectively. [110] addresses the problem with a notion called the disambiguability of faulty traces. In the context of this thesis, the disambiguability of faulty traces reflect the determinism of the faulty plant $\mathcal{N}_p^f$. The authors claim that the disambiguability of faulty traces has to decrease for the reconfigurability to increase. With other words, the faulty plant has to be less nondeterministic or more deterministic for the control loop to become more reconfigurable.

This section presented a selection of works dealing with fault-tolerant control of discrete-event systems. A broader overview on other approaches and related problems is given in [13]. In the latter reference, the main ideas found in literature are explained in the nomenclature of [126] which is also used in this thesis in order to simplify the comparison of approaches and results.

1.4.3. Comparative study of literature

This section discusses some concepts and properties taken from the literature and stresses the necessary extensions or modifications added in this thesis.

Model properties. The approach of this thesis differs from the existing ones firstly by the model class, namely I/O automata. This modeling approach is motivated by the fact that I/O automata explicitly describe the action-reaction principle (causality) which is a fundamental property of technological systems. The control problem has been solved for standard automata through the Supervisory Control Theory (SCT). One idea would be to adopt the existing relation between I/O automata and standard automata (see [127]) in order to deduce an I/O controller from the SCT-based supervisor. A similar approach has been proposed by [149, 150] for DES with partial observations. The resulting I/O controller is a Moore automaton. In the framework depicted in Fig. 1.1, this way would lead to an open-loop control structure because the control signals w_c would be generated regardless of the outputs w_p of the plant but depending only on the internal state z_c of the controller due to the Moore property.

The system considered in the references above are mostly of deterministic nature with a Moore-property. In this thesis, nondeterministic automata are used to model the behavior of the plant. The notion of stable and unstable state (called transient here) is omitted here because of the self-loops at every stable state and the unnecessarily induced long events sequences along unstable states which do not reflect the signal produced by a real hardware. In addition, this assumption is inadequate for systems where any input leads to another different state and the plant waits until then, meanwhile ignoring the last output. In this thesis, such self-loops are not considered at all because an explicit illustration of the ignorance of an input by a self-loop will unnecessarily augment the complexity of the plant, if this were to be done at every state.

Specification of the process. A main difference between literature and this thesis is the definition of the specification for the nominal plant and the faulty case. In literature, several specifications are usually defined with decreasing restrictions depending on the severity of the fault. In this thesis, the specification is the same for the nominal and the faulty case. However, the specification used here for fault tolerance already contain degrees of freedom which are not necessary in the nominal case but might be useful in case of a fault. In addition, the model of the nominal and the model of the faulty plant are all in one overall model which differs from the approach presented here. Instead, this thesis deals with models of the nominal and the faulty plant separately without mixing them together.

Most approaches based on the supervisory control of standard automata like [110, 144] and [186] separate the specification of the nominal system from the specification of the faulty system. The former is usually a degraded version of the latter. On the contrary, the specification defined in this thesis is not degraded for control reconfiguration nor modified but maintained for the faulty plant. However, degrees of freedom are foreseen in the specification formalism to enable some flexibility in the process towards certain faults.

Control design. The control design method used in this thesis only considers controllable events since they are the only relevant ones to achieve a given specification. Several types of specifications which can strictly be based on the single states, inputs, outputs, sequences or combinations are addressed in this thesis, instead of language based specifications of the literature. Faults are embedded in the specification modeling, thus, permit to formalize the degradation of specifications. An appropriate definition of the notion of redundancy degree instead of tolerance relations used in tolerance automata is proposed here. The reconfiguration steps are tackled based on the assumption that the plant is safely diagnosable and safely controllable in the sense of [144]. The crucial difference with the reconfiguration method presented here is the fact that, the reconfiguration approach proposed here enables to solely modify the transitions of the control law automaton while its structure remains unchanged, instead of multiplexing among a predefined bank of controllers.

Degree of fault tolerance and control reconfiguration. In [173] the redundancy degree reflects the number of actuators and sensors available to guarantee the fulfillment of "keeping" the controllability and the observability of the plant. The approach is based on a graph which requires 2^N states for a system with N components. In this thesis, the redundancy degree is the number of alternative state transitions or output sequences in a controller from an initial state to a given state which do not violate the specification.

Another difference with the subsequent approach is that in this thesis a nominal controller steers a nondeterministic plant until a fault is diagnosed. Starting from the faulty state of the plant, the reconfiguration unit adapts the existing controller to the fault.

Implementation of methods for discrete-event systems. The main issues to consider in software development for discrete-event systems are extensively discussed in [182]. A possible way to implement models of a plant is to use binary decision diagrams (BDDs) developed by [117] made prominent in [28, 53]. They have already been applied for supervisory control after [156] by [95], for the optimal control problem by [178] and efficient search algorithms [102]. Instead of BDDs, I/O automata modeling a plant or a controller in this thesis are implemented by tables which represent a list of transitions from a state z, to a next state z with the output w for the input v. Since the characteristic function $L(\cdot)$ used here is a binary function, it is possible to express it as a BDD but the implementation is not the topic of this thesis.

Most software tools for discrete-event systems computation and analysis are dedicated to standard automata. However, the functionalities and programming method for standard automata can be easily extended to the I/O automata used in this thesis. A selection of existing software tools which support the supervisor synthesis consists of TCT [176], DESUMA [163], Supremica [29] or DESLAB [59] recently presented. SIGALI [131] is a tool for implementation of discrete-event controllers. The simulation and experimental results presented in this thesis were obtained with MATLAB/Simulink programs dedicated to I/O automata and the methods developed here. Appendix C gives an overview of the functions used here.

The issue of deriving, e.g., a PLC code from a supervisor or a controller is an important issue of practical relevance but not discussed here. Instead, the PLCs used in the examples of this thesis solely exchange the commands and sensor values as an interface module between the plant and the controller implemented in MATLAB/Simulink. However, an approach to derive a PLC code out of a supervisor is investigated in [124] and some verification method are given in [44, 83, 111] and [157].

1.5. Main results of the thesis

1.5.1. Formalism for fault-tolerant control of I/O automata

In order to propose a formal and systematic solution to the fault-tolerant control problem of I/O automata, it is necessary to setup a mathematical framework enabling the problem to be addressed in a uniformed way. The mathematical foundation is inspired from [97, 126] and [56] whereas the nomenclature is mostly from [126]. Further extensions are necessary to deal with the nondeterminism of plants, degrees of freedom of a specification, its feasibility, redundancy degree of controllers, etc. The main extensions, modifications and particularities developed in this thesis are:

- **Active input, active output and active next state operators.** These operators permit to precisely select a set of inputs, outputs or states depending on the needs (see Section 2.2.1). For instance, the active output operator permits to read only output signals w which can be generated from a given state z for a particular input v. In general, a variety of input parameters is possible, such as single states, states combinations, state-input combination, input-output combination, etc. Their applicability is effective in control design [2], fault modeling [3] and control reconfiguration [1, 4–6].

- **Appropriate use of symbolic adjacency matrices.** Numerical and symbolic adjacency matrices are used in correspondingly defined rings and subrings (see Section 2.2.3). Thus, mathematical foundations are proposed to enable basic operations as the addition and the multiplication. Their interpretation is not limited to reachability analysis but adapted to I/O automata in order to explicitly read signal sequences, compute their length or formalize searching algorithms. This rigorous background is used in [15] for a formal control design and in [8] for a formal control reconfiguration.

- **I/O trellis automata.** Alternatively to static automata graphs, trellis offer an intuitive dynamic representation of discrete-event processes. They are formalized in Section 2.2.5 to fit in the I/O automata framework. Cyclic and noncyclic state sequences are better reflected by trellis especially when degrees of freedom or redundancies need to be visualized. Their applicability for fault-tolerant control is demonstrated in [5].

- **Special properties of the control loop.** The notion of *output-determinism* of a system, also called W-*determinism,* is introduced in Section 2.2.4 to handle control loops where the plant might be nondeterministic but the control output generation of the controller is deterministic. This notion was previously introduced in [1] and used in [2] to tackle the algebraic loop during the control design. The concept of weak well-posedness which solves the algebraic loop was first developed in [10]. Moreover, criteria are proposed to analyze the well-posedness of a control loop. Since the control loop usually blocks when there is a fault, a relation is established between the nonblockingness of the control loop and its well-posedness. This property is essential to study the reconfigurability of a control loop.

The formal extensions mentioned above are used in definitions, proofs and the formulation of important properties like the feasibility of a specification, the controllability of a plant, the formal representation of a fault and the reconfigurability of a controller for a given fault and a specification.

1.5.2. Modeling of faults and failures

This contribution is motivated by the fact that a successful control reconfiguration is only possible with a knowledge of the faulty behavior of the plant. The trajectory of the faulty system is supposed to be unknown. Thus, the method developed in [3] to obtain the model of the faulty plant subject to actuator, sensor and system internal faults and failures is implemented by the reconfiguration unit in order to forecast the behavior of the faulty plant and eventually anticipate with the appropriate correcting actions.

The idea is to introduce **error relations** to model an input, an output or a state transition falsification as explained in Section 3.5. Depending on the signal falsification, faults and failures at the actuator, sensor and system internal level are modeled:

- Faults are modeled by *changes of signals* from the original ones but which are still in the set of inputs, outputs or states originally defined. For example, the closing command of a valve can be falsified into an opening command so that the valve opens when the controller tries to close it.

- Failures represent signal falsification which are *out of the scope* of the considered signal set. Therefore, they are modeled by the ε symbol. It can be used to model a broken sensor which returns a voltage which can not be interpreted by the controller.

In Section 3.6, the error relations and the active input, output, and next state operators are used with the characteristic function of the nominal plant $\mathcal{N}_p$ to deduce the characteristic function of the faulty plant $\mathcal{N}_p^f$. The latter and the specification $\mathcal{S}$ are used by the reconfiguration unit to adapt the controller accordingly.

1.5.3. Nominal controller design method

The control design problem solved in this thesis consists of a plant $\mathcal{N}_p$ and a specification $\mathcal{S}$. $\mathcal{N}_p$ is a nondeterministic I/O automaton which symbolizes the free behavior of the plant without control and without any fault. The specification $\mathcal{S}$ symbolizes the goal to be achieved by the plant under the influence of the controller $\mathcal{A}_c$. Possible goals are *a target state to reach, a state sequence or an output sequence to follow*. This specification implicitly involves specifications with forbidden states or output sequences. The control method presented in Section 4.3 aims at finding a control law $\mathcal{A}_c$ which enforces the fulfillment of the specification $\mathcal{S}$ in the plant $\mathcal{N}_p$. The main steps are the following:

- **Build the specification automaton $\mathcal{N}_s$** out of the model of the plant $\mathcal{N}_p$ and the specification $\mathcal{S}$. The result is an I/O automaton which contains only those transitions from $\mathcal{N}_p$ where $\mathcal{S}$ can be achieved. Every transition which violates $\mathcal{S}$ is deleted from $\mathcal{N}_p$ to obtain $\mathcal{N}_s$. The remaining transitions are those which are activated by the controller in the closed-loop. Formal methods are suggested in terms of homomorphism in [1] and signal sequences. The simple feasibility of a specification in a plant $\mathcal{N}_p$ does not guarantee its fulfillment but solely the possibility of its fulfillment. For the guarantee of fulfillment, the notion of safe feasibility is covered by Theorem 4.1, its Corollary

4.1 and Theorem 4.2 for different specifications. It is particularly helpful for nondeterministic plants $\mathcal{N}_p$ where the required behavior overlaps with the forbidden one, i.e, when they share some transitions.

- **Reverse the I/O behavior of** $\mathcal{N}_s$ to obtain $\mathcal{N}_c$, the maximally permissive controller. Since $\mathcal{N}_c$ might be nondeterministic two control law synthesis are presented to

  1. directly synthesize a deterministic control law $\mathcal{A}_c$ or

  2. decompose the $\mathcal{N}_c$ in deterministic control laws $\mathcal{A}_c$.

Once the controller $\mathcal{A}_c$ is computed, it needs to be validated through simulations and experiments. Hence, it is necessary to have a fixed structure where the computed control law just need to be loaded in order to run the control loop. For this sake, *a novel realization scheme of the discrete-event feedback controller* is presented in this thesis (see Section 4.3.3). A rough architecture is proposed in [15] and a detailed version is proposed in [2].

The *controllability* of a plant is proven in Theorem 4.4 to be given by the existence of an output-deterministic controller $\mathcal{A}_c$ which guarantees the fulfillment of a specification and the weak well-posedness of the control loop [2].

1.5.4. Off-line and On-line reconfiguration method

This section introduces both reconfiguration techniques developed in this thesis: off-line and on-line reconfiguration. The main difference between both methods is the scope in which they can be applied. Off-line reconfiguration consists of a global adaptation of the control law whereas on-line reconfiguration consists only of a local adaption. A global adaptation involves every possible transition of the control law but a local adaptation focuses only on the currently faulty transition. Therefore, it is suitable to realize a global adaptation *off-line* and a local adaptation *on-line*. However, both methods share the following main steps of control reconfiguration:

1. Build the model of the plant under the influence of faults and/or failures

2. Choose the appropriate strategy between forward and backward recovery

3. Execute the controller reconfiguration for the following situations if possible:

   - Actuator faults and failures: Input/Output replacement.

   - System internal faults: path replacement.

   - Sensor faults and failures: output replacement.

Now, the two reconfiguration methods are explained.

Off-line reconfiguration method. Control reconfiguration steps which are applicable off-line are classified in two techniques published in [1, 5] and explained in Section 5.1.2 of this document:

- **Trajectory re-planning** which consists of modifying the nominal state sequence to be enforced by the controller according to the specification. The input and output signals of the new control law are thereafter adapted to the new state sequence.

- **Input/Output adaptation** which focuses solely on modifying the input and output signals of the control law according to the fault at hand. A result worth mentioning is the fact that *redundancies are not always required* to achieve fault tolerance in cases where the knowledge on actuator or sensor signal falsification is sufficiently available. Sufficient here means that a signal falsification map can be established between nominal and faulty signals.

These two techniques are applied on the whole controller, i.e, every I/O transition is eventually adapted to the fault.

On-line reconfiguration method. In this case, the main reconfiguration techniques are applied only for the blocking transition as presented in [4, 6, 7] and [8], instead of every transition as in the off-line reconfiguration case. Fig. 1.4 shows the possible adaptations of the state trajectory of the controller and illustrates the consequences on the plant depending on the selected reconfiguration strategy (see Section 5.3). The main reconfiguration strategies are:

- **Backward (BW) recovery:** If the fault is diagnosed to have occurred at step k_f in the past, the controller forces the plant to move from the current state $Z_p(k)$ to the last healthy one $Z_p(k_f)$ with $k_f < k$.

- **Forward (FW) recovery:** The controller forces the plant to move from the state $Z_p(k_f)$ where the fault occurred either to the specified next state with a redundant control signal or to another next state which is different from the previous one but still satisfies the specification.

- **Repair procedure:** The dynamic of the plant is first "frozen" to avoid dangerous states or transitions. The model of the faulty plant $\mathcal{N}_p^f$ is computed. Then the reconfiguration unit decides whether forward or backward recovery should be started.

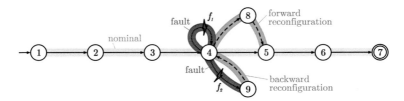

Figure 1.4.: Illustration of three main behaviors of a plant: nominal, faulty and reconfigured

According to the chosen reconfiguration strategy, the dashed line in Fig. 1.1 is activated to modify specific components of the controller $\mathcal{A}_c$. This approach uses the feedback controller realization architecture developed in [2]. It includes, among other components, a redundance-free control law $\mathcal{A}_c$ and a state sequence generator $\overline{Z}_s(0 \ldots k_e)$. $\mathcal{A}_c$ is the nominal controller. That is the one which is used before the fault occurs. $\overline{Z}_s(0 \ldots k_e)$ consists of states which have to be followed by the plant at each step k_s which is the internal counter of the nominal controller $\mathcal{A}_c$.

The method presented here alters the components cited above by the following steps when a fault occurs:

1. Computation of a correcting term k_c which modifies k_s.

2. Update of the state sequence generator $\overline{Z}_s(0 \ldots k_e)$ through an appropriate subsequence $\overline{Z}_s(k_{sf} \ldots k_{sf} + l)$ where k_{sf} is the value of the internal counter of the controller when the fault is known to have occurred and l is the number of necessary steps needed to reach the next healthy state.

3. Adapt the nominal control law $\mathcal{A}_c$ by deleting or adding transitions where the characteristic function $L_c(\cdot)$ of $\mathcal{A}_c$ is set to 0 or 1 correspondingly. $\mathcal{A}_c^r$ then becomes the reconfigured control law.

The reconfiguration algorithm which implements this method has a polynomial complexity. Thus, the fault-tolerant control problem handled here is tractable because a solution can always be found in finite time, if it exists.

Reconfigurability. The formal definition of the reconfigurability presented in Section 5.4 is to recover controllability of the faulty plant by a new controller. This incorporates the nonblockingness of the reconfigured control loop. The controller of a faulty plant that guarantees the fulfillment of the original specification and the nonblockingness of the closed-loop, hence, the controllability, is said to be a *fault-tolerant controller*.

1.5.5. Simulation and experimental results

The applicability of the concepts developed in this thesis have been investigated for two categories of systems:

- **Academic examples**: the fluid level control and the virtual manufacturing process of Section 1.3.1 have been used for simulations.

- **Real plants**: a 3-Tank system (Section 4.4.1), a pilot manufacturing cell (Chapter 7) have been used for experiments at the Institute of Automation and Computer Control at the Ruhr-Universität Bochum. A pilot filling line of the company Siemens AG served as a benchmark process to compare fault-tolerant control methodologies of discrete-event systems.

These examples confirmed that the reconfigurability of a controller for a faulty plant depends on the feasibility of the specification in the faulty plant. The main limitation encountered was the system complexity which is inherent in the discrete-event model. However, the tractability of the reconfiguration problem has been observed in the pilot manufacturing cell for a system internal fault (see Section 7.5).

1.6. Structure of the thesis

Part I deals with the formal results of this thesis which are classified in four categories from Chapter 2 to Chapter 5. The key topics are modeling of discrete-event systems in general, control design and control reconfiguration.

Chapter 2 presents the basic foundations required to interpret the results developed in this thesis. First, nondeterministic I/O automata are then presented as the fundamental model of the class of system which are the focus here. Formal tools and properties developed for the resolution of the fault-tolerant control problem are explained in detail. Some modeling issues of automata theory are put in the context of fault-tolerant control with nondeterministic I/O automata.

Chapter 3 represents the first step towards reconfiguration once a fault is detected and identified in the system. The error relations are introduced to solve the fault modeling problem for actuator, sensor and system internal faults and failures.

Chapter 4 is devoted to the nominal control design problem which precedes the fault-tolerant control problem. Since the control design method is novel, important assumptions and fundamental questions regarding the nominal control loop are stated for clarity. After an introduction on the concept of specification, the control design method is explained in Section 4.3. The applicability of the approach is demonstrated on a 3-Tank system.

Chapter 5 deals with the main topic of this thesis. It contains the theoretical objective pursued by previous chapters. Detailed explanations are given on off-line and on-line control reconfiguration. Degrees of freedom and redundancies which are crucial for control reconfiguration are formalized in the nondeterministic I/O automata framework. The reconfigurability conditions follow as another important result of this thesis. A complexity analysis is provided to show the tractability of the fault-tolerant control problem. The off-line control reconfiguration of the fluid level control process is presented.

Part II contains demonstrations of the methods developed in Part I concerning modeling in the nominal and faulty case, control design and control reconfiguration. The complexity of the examples increases from Chapter 6 to Chapter 7.

Chapter 6 focuses on the demonstration of the on-line control reconfiguration method applied on a virtual manufacturing process. The forward control reconfiguration is applied for sudden random workpiece supply. The backward control reconfiguration is applied after a sensor fault.

Chapter 7 demonstrates the practical results obtained by applying the control reconfiguration methods on a pilot manufacturing cell which is the example with the highest complexity here. This example shows how the reconfiguration solution can be found and executed by a computer with algorithms implemented in MATLAB/Simulink.

Chapter 8 gives a summary of the thesis. The most relevant issues are presented and open problems are discussed with sketches of solutions. The latter can be developed in future works.

In **Part III**, appendices are given to ease the reading flow throughout the thesis and to enable a deep understanding of the results.

Appendix A summarizes the symbols used throughout the thesis.

Appendix B contains the proofs of all lemmas and theorems.

Appendix C is a description of the MATLAB/Simulink toolbox IDEFICS developed for the purposes of discrete-event system modeling, control design, control loop simulation and simulations of fault-tolerant control.

Appendix D is a compilation of all components of the pilot manufacturing cell. The symbols representing input events, output events and states are given in tables. These tables are necessary to understand the results of Chapter 7 and interpret them correctly.

Part I.

Theory of discrete-event systems for fault-tolerant control

2. Modeling and analysis methods for discrete-event systems

Abstract. *All automata models used in this thesis are introduced in this chapter. Active set operators are defined in order to compute the output and transition degree of I/O automata graphs. The latter is used in numerical adjacency matrices to compute the number of possible paths among given states. The symbolic adjacency matrices presented in this chapter reveal the explicit state, input or output sequences between two states. The deterministic output generation of an automaton is formalized by the W-determinism property which the controller has to ensure as required in Chapter 4.*

2.1. Description of discrete-event systems

2.1.1. Autonomous automata

Definition 2.1 (Autonomous automaton). *An autonomous automaton $\mathcal{M}$ has no events and is defined as*

$$\mathcal{M} = (\mathcal{Z}, \lambda, z_0) \tag{2.1}$$

with $\mathcal{Z}$ as the set of states, λ as the state transition function and z_0 as the initial state.

$$\lambda : \mathcal{Z} \times \mathcal{Z} \to \{0, 1\} \text{ with } \lambda(z', z) = \begin{cases} 1, & \text{if } (z', z)! \\ 0, & \text{else,} \end{cases} \tag{2.2}$$

where $(z', z)!$ means that the system can carry out a state transition from z to z'.

2.1.2. Nondeterministic I/O automata

Definition 2.2 (Nondeterministic I/O automaton). *A nondeterministic I/O automaton $\mathcal{N}$ is defined as a tuple*

$$\mathcal{N} = (\mathcal{Z}, \mathcal{V}, \mathcal{W}, L, z_0), \tag{2.3}$$

with the following additional elements: $\mathcal{V}$ the set of control inputs, $\mathcal{W}$ the set of control outputs and L the characteristic function. The dynamics of the automaton is given by the function

$$L : \mathcal{Z} \times \mathcal{W} \times \mathcal{Z} \times \mathcal{V} \to \{0, 1\}$$

$$L(z', w, z, v) = \begin{cases} 1, & \text{if } (z', w, z, v)! \\ 0, & \text{else,} \end{cases} \tag{2.4}$$

where $(z', w, z, v)!$ means that the system $\mathcal{N}$ can move from state z with the input v to state z' and generate the output w.

$Z(0 \cdots k_e)$ represents a state sequence of the length $k_e + 1$ with the elements $Z(k)$, $k = \ldots k_e$. For a given set $\boldsymbol{K_e} = \{k_{e1}, \ldots, k_{e\kappa}\}$ of horizons, $\mathcal{Z}(0 \cdots \boldsymbol{K_e})$ is the set of state sequences $Z_i(0 \cdots k_{ei})$. Thus,

$$\mathcal{Z}(0 \cdots \boldsymbol{K_e}) = \{Z_1(0 \cdots k_{e1}), \ldots, Z_\kappa(0 \cdots k_{e\kappa})\}. \tag{2.5}$$

If $k_{e1} = k_{e2} = \ldots = k_{e\kappa} = k_e$, then all corresponding state sequences have the same length $k_e + 1$. An infinite repetition of a state sequence is characterized by the $*$ symbol as $Z^*(0 \cdots k_e) = (Z(0), \ldots, Z(k_e))^* = (Z(0), \ldots, Z(k_e), Z(0), \ldots)$. The $\wedge$ and $\vee$ symbols represent the Boolean AND and OR operations whereas $\bigwedge$ and $\bigvee$ symbolize their recursive application. In the latter case, the upper limit can be a maximal value of an index or a set of values recursively taken by the variable of the lower limit. Since the characteristic functions L and λ can only have the value 1 or 0, they will be used sometimes with both Boolean and arithmetic operators like $\sum$ and $\prod$.

Definition 2.3 (Subautomata and superautomata). *A given I/O automaton $\mathcal{N}_2 = (\mathcal{Z}_2, \mathcal{V}_2, \mathcal{W}_2, L_2, \mathcal{Z}_{02})$ is a subautomaton of an I/O automaton $\mathcal{N}_1 = (\mathcal{Z}_1, \mathcal{V}_1, \mathcal{W}_1, L_1, \mathcal{Z}_{01})$ if*

$$\mathcal{Z}_2 \subseteq \mathcal{Z}_1, \mathcal{V}_2 \subseteq \mathcal{V}_1, \mathcal{W}_2 \subseteq \mathcal{W}_1$$

hold and L_2 is a restriction of L_1 to the set $\mathcal{Z}_2 \times \mathcal{W}_2 \times \mathcal{Z}_2 \times \mathcal{V}_2$ in the sense of [81]. The restriction means that the behavior of $\mathcal{N}_2$ is included in the one of $\mathcal{N}_1$, so that

$$\mathcal{N}_2 \subseteq \mathcal{N}_1$$

holds. On the basis of [92], $\mathcal{N}_1$ is said to be a superautomaton of $\mathcal{N}_2$.

Completeness. The completeness property of nondeterministic I/O automata is needed for model-based diagnosis to avoid that a fault is wrongly excluded from the diagnosis result, thus, insures that no fault has been missed.

Definition 2.4 (Completeness). *A nondeterministic I/O automaton $\mathcal{N}$ is said to be complete if $\forall (z, v) \in \mathcal{Z} \times \mathcal{V}, \exists (z', w) \in \mathcal{Z} \times \mathcal{W} : L(z', w, z, v) = 1$.*

A plant $\mathcal{A}_p$ modeled by a deterministic and incomplete I/O automaton is exemplary depicted in Fig. 2.1.

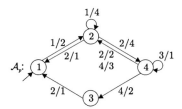

Figure 2.1.: Example of a plant modeled as a deterministic and incomplete I/O automaton

2.1.3. Autonomous behavior of a system

The autonomous behavior of the plant is to be defined here in a way that it can be referred to while describing the "freezing" procedure of the plant before the reconfiguration starts, namely when the controller generates $w_c = \varepsilon$. It should be clear that if $w_c = \varepsilon$ coincides with an explicit command signal $v_p = \varepsilon$ which may have been used or designed for other purposes, the automaton may continue to update its states due to the own dynamic of the plant. That means that some measurements may still be sent to the controller which continue steering the plant accordingly to its control policy.

Definition 2.5 (Autonomous behavior). *The autonomous behavior of an automaton is the set of executed state sequences $Z(0 \cdots k_e + 1)$ and generated output sequences $W(0 \cdots k_e)$*

by the automaton for the input sequence $V(0 \cdots k_e) = \varepsilon \ldots \varepsilon$ with $k_e \in \mathbb{N}$, where $L(Z(k+1), W(k), Z(k), \varepsilon) = 1$ for $k = 1 \ldots k_e$.

2.1.4. Asynchronous networks of I/O automata

For component-oriented modeling issues the subsystems are described by deterministic I/O automata extended with coupling signals in order to mirror the interactions among the components [66, 127]. An *extended I/O automaton* is defined as

$$\mathcal{N}_\iota = (\mathcal{Z}_\iota, \mathcal{V}_\iota, \mathcal{W}_\iota, \mathcal{N}_{S\iota}, \mathcal{N}_{R\iota}, F_\iota, G_\iota, H_\iota, z_{0_\iota}), \tag{2.6}$$

where each element has the following meaning for every ι-th component

$$
\begin{aligned}
\mathcal{Z}_\iota &\quad - \quad \text{State set} \\
\mathcal{V}_\iota &\quad - \quad \text{Control input set} \\
\mathcal{W}_\iota &\quad - \quad \text{Control output set} \\
\mathcal{N}_{S\iota} &\quad - \quad \text{Interconnection inputs set} \\
\mathcal{N}_{R\iota} &\quad - \quad \text{Interconnection outputs set} \\
F_\iota &\quad - \quad \text{Interconnection output function} \\
G_\iota &\quad - \quad \text{State transition function} \\
H_\iota &\quad - \quad \text{Output transition function} \\
z_{0_\iota} &\quad - \quad \text{Initial state.}
\end{aligned}
$$

The cardinal numbers of the additional sets above are

$$
\begin{aligned}
|\mathcal{N}_{S\iota}| &= T \\
|\mathcal{N}_{R\iota}| &= P.
\end{aligned}
$$

The functions F_ι, G_ι and H_ι are defined as

$$
\begin{aligned}
F_\iota &: \mathcal{Z}_\iota \times \mathcal{V}_\iota \times \mathcal{N}_{Si} \to \mathcal{N}_{R\iota} \\
r_\iota &= F_\iota(z_\iota, v_\iota, s_\iota),
\end{aligned} \tag{2.7}
$$

$$G_\iota : \mathcal{Z}_\iota \times \mathcal{V}_\iota \times \mathcal{N}_{S\iota} \to \mathcal{Z}_\iota$$
$$z'_\iota = G_\iota(z_\iota, v_\iota, s_\iota), \tag{2.8}$$

$$\text{and } H_\iota : \mathcal{Z}_\iota \times \mathcal{V}_\iota \times \mathcal{N}_{S\iota} \to \mathcal{W}_\iota$$
$$w_\iota = H_\iota(z_\iota, v_\iota, s_\iota). \tag{2.9}$$

A tuple $(z_\iota, v_\iota, s_\iota)$ represents the situation where the extended automaton $\mathcal{N}_\iota$ is in the state z_ι getting the control input v_ι and the interconnection input s_ι. The extended automaton responds with the tuple $(z'_\iota, w_\iota, r_\iota)$ by moving to the state z'_ι while generating the control output w_ι and the interconnection output r_ι.

The overall state of the system and its interaction with its environment are represented by the signal vectors

$$\boldsymbol{z} = (z_1, \cdots, z_\mu)^T \in \boldsymbol{\mathcal{Z}} = \mathcal{Z}_1 \times \cdots \times \mathcal{Z}_\mu \tag{2.10}$$
$$\boldsymbol{v} = (v_1, \cdots, v_\mu)^T \in \boldsymbol{\mathcal{V}} = \mathcal{V}_1 \times \cdots \times \mathcal{V}_\mu \tag{2.11}$$
$$\boldsymbol{w} = (w_1, \cdots, w_\mu)^T \in \boldsymbol{\mathcal{W}} = \mathcal{W}_1 \times \cdots \times \mathcal{W}_\mu \tag{2.12}$$

whereas the interaction among the components is explicitly described by the interconnection signal vectors

$$\boldsymbol{s} = (s_1, \cdots, s_\mu)^T \in \boldsymbol{\mathcal{N}_S} = \mathcal{N}_{S1} \times \cdots \times \mathcal{N}_{S\mu} \tag{2.13}$$
$$\boldsymbol{r} = (r_1, \cdots, r_\mu)^T \in \boldsymbol{\mathcal{N}_R} = \mathcal{N}_{R1} \times \cdots \times \mathcal{N}_{R\mu}. \tag{2.14}$$

Interconnection signals are linked to each other by the interaction block $\boldsymbol{K}$ (Fig. 2.2). The coupling block $\boldsymbol{K}$ describes the structure of the connection between the components of the system. This is modeled by a square matrix with the dimension $(\mu \times \mu)$ where

$$\boldsymbol{s} = (s_1, \cdots, s_\mu)^T = \boldsymbol{K} \cdot (r_1, \cdots, r_\mu)^T = \boldsymbol{K} \cdot \boldsymbol{r} \tag{2.15}$$

with

$$\boldsymbol{K} = (k_{ij}) \text{ with } k_{ij} = \begin{cases} 1, & \text{if } s_i = r_j \text{ holds} \\ 0, & \text{else.} \end{cases} \tag{2.16}$$

Without the restriction (2.16), having more ones in a row would lead to the addition of discrete-event values or symbols. Such an operation is not defined and therefore, has to be excluded.

Equation (2.16) implies that K can be written as

$$K = \begin{pmatrix} k_1^T \\ \vdots \\ k_\mu^T \end{pmatrix}. \tag{2.17}$$

For the example of Fig. 2.2 the relation between the interconnection signals is given by

$$\begin{pmatrix} s_1 \\ s_2 \\ s_3 \end{pmatrix} = \begin{pmatrix} 0 & 0 & 1 \\ 1 & 0 & 0 \\ 0 & 1 & 0 \end{pmatrix} \cdot \begin{pmatrix} r_1 \\ r_2 \\ r_3 \end{pmatrix}.$$

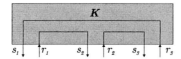

Figure 2.2.: Example of a coupling block

In summary, the I/O automata network $\mathcal{AN}$ is defined by

$$\mathcal{AN} = (\{\mathcal{N}_1, \cdots, \mathcal{N}_\mu\}, \mathcal{Z}, \mathcal{V}, \mathcal{W}, \mathcal{N}_S, \mathcal{N}_R, K, z_0)$$
$$\text{with } z_0 = (z_{0_1}, \cdots, z_{0_\mu})^T \tag{2.18}$$

as depicted in Fig. 2.3.

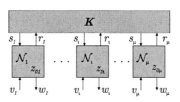

Figure 2.3.: Asynchronous I/O automata network

The graph of an extended I/O automaton $\mathcal{N}_\iota$ is exemplary depicted in Fig. 2.4 with each transition labeled by a $v/w/s/r$ tuple.

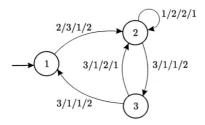

Figure 2.4.: Example of the graph of an extended deterministic I/O automaton

The overall interconnection function $\boldsymbol{F}$ involves the interconnection function F_ι of each extended I/O automaton $\mathcal{N}_\iota$ in the network:

$$\boldsymbol{r} = \boldsymbol{F}(\boldsymbol{z}, \boldsymbol{v}, \boldsymbol{s}) = \begin{pmatrix} F_1(z_1, v_1, s_1) \\ \vdots \\ F_\mu(z_\mu, v_\mu, s_\mu) \end{pmatrix}. \tag{2.19}$$

The overall functions $\boldsymbol{G}$ and $\boldsymbol{H}$ are defined analogously by

$$\boldsymbol{z}' = \boldsymbol{G}(\boldsymbol{z}, \boldsymbol{v}, \boldsymbol{s}) = \begin{pmatrix} G_1(z_1, v_1, s_1) \\ \vdots \\ G_\mu(z_\mu, v_\mu, s_\mu) \end{pmatrix} \quad \text{and} \tag{2.20}$$

$$\boldsymbol{w} = \boldsymbol{H}(\boldsymbol{z}, \boldsymbol{v}, \boldsymbol{s}) = \begin{pmatrix} H_1(z_1, v_1, s_1) \\ \vdots \\ H_\mu(z_\mu, v_\mu, s_\mu) \end{pmatrix}. \tag{2.21}$$

If an automata network $\mathcal{AN}$ is well-posed its composition leads to an equivalent single automaton $\mathcal{A}$ defined as

$$\mathcal{A} = (\mathcal{Z}, \mathcal{V}, \mathcal{W}, \tilde{\boldsymbol{G}}, \tilde{\boldsymbol{H}}, z_0) \tag{2.22}$$

with

$$\tilde{G}(z, v) = \begin{pmatrix} G_1(z_1, v_1, \mathbf{k}_1^T \cdot \tilde{r}) \\ \vdots \\ G_\mu(z_\mu, v_\mu, \mathbf{k}_\mu^T \cdot \tilde{r}) \end{pmatrix},$$

$$(2.23)$$

$$\tilde{H}(z, v) = \begin{pmatrix} H_1(z_1, v_1, \mathbf{k}_1^T \cdot \tilde{r}) \\ \vdots \\ H_\mu(z_\mu, v_\mu, \mathbf{k}_\mu^T \cdot \tilde{r}) \end{pmatrix}$$

where

$$\tilde{r} \in \tilde{\mathcal{N}}_r = \{\tilde{r} \in \mathcal{N}_{R1} \times \cdots \times \mathcal{N}_{R\mu} : \tilde{r} = F(z, v, K \cdot \tilde{r})\} \qquad (2.24)$$

with z_0 defined as in (2.18). Note that the interconnection F does not appear in (2.22) in contrast to (2.6) because there exists no interconnection signals after the components have been coupled together to form a single I/O-automaton. The focus is set on the input/output behavior of the composed automaton (Fig. 2.5). In addition, (2.23) can be simplified using (2.20) and (2.21) to

$$\tilde{G}(z, v) = G(z, v, K \cdot \tilde{r})$$
$$\tilde{H}(z, v) = H(z, v, K \cdot \tilde{r}) \qquad (2.25)$$

which clearly shows that a vector $\tilde{r}$ is not visible from outside the system. Note that (2.25) holds only if the I/O automata network is well-posed. The well-posedness of a nondeterministic asynchronous network of extended I/O automata is extensively discussed in [10].

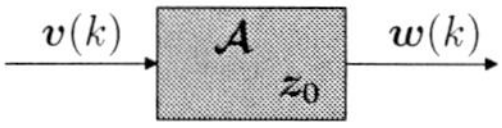

Figure 2.5.: Composed I/O-automaton

The resulting dynamic behavior of the composed I/O-automaton $\mathcal{A}$ is represented by

$$\boxed{\begin{array}{l} z(k + 1) = \tilde{G}(z(k), v(k)), \quad z(0) = z_0 \\ w(k) = \tilde{H}(z(k), v(k)). \end{array}}$$

$$(2.26)$$

2.2. Analysis tools for discrete-event fault-tolerant control

2.2.1. Active input, active output and active next state sets

$\mathcal{V}_a(\cdot)$, $\mathcal{W}_a(\cdot)$ and $\mathcal{Z}_a(\cdot)$ are introduced as the *active input, active output and active next state sets*, respectively, of a given state z, a state couple (z', z), a state-output couple (z, w), or a state-input (z, v) couple in an I/O automaton:

$$\mathcal{V}_a(z) = \{v \in \mathcal{V} : \sum_{z'}^{z} \sum_{w}^{w} L(z', w, z, v) > 0\} \tag{2.27}$$

$$\mathcal{V}_a(z', z) = \{v \in \mathcal{V} : \sum_{w}^{w} L(z', w, z, v) > 0\} \tag{2.28}$$

$$\mathcal{V}_a(z, w) = \{v \in \mathcal{V} : \sum_{z'}^{z} \sum_{z}^{z} L(z', w, z, v) > 0\} \tag{2.29}$$

$$\mathcal{W}_a(z) = \{w \in \mathcal{W} : \sum_{z'}^{z} \sum_{v}^{v} L(z', w, z, v) > 0\} \tag{2.30}$$

$$\mathcal{W}_a(z', z) = \{w \in \mathcal{W} : \sum_{v}^{v} L(z', w, z, v) > 0\} \tag{2.31}$$

$$\mathcal{W}_a(z, v) = \{w \in \mathcal{W} : \sum_{z'}^{z} L(z', w, z, v) > 0\} \tag{2.32}$$

$$\mathcal{Z}_a(z) = \{z' \in \mathcal{Z} : \sum_{w}^{w} \sum_{v}^{v} L(z', w, z, v) > 0\} \tag{2.33}$$

$$\mathcal{Z}_a(z, v) = \{z' \in \mathcal{Z} : \sum_{w}^{w} L(z', w, z, v) > 0\}. \tag{2.34}$$

According to the context, the subscript "x" of $\mathcal{N}_x$ will be replaced by "c" for the controller, "s" for the specification or "p" for the plant. For example, $\mathcal{V}_{ac}(z_p)$ denotes the active input set from state z_p in $\mathcal{N}_c$.

Figure 2.6 shows the internal structure of a nondeterministic I/O automaton $\mathcal{N}_p$ where the SR block is a shift register initialized with some $z_{0p} \in \mathcal{Z}_{0p}$. It saves the computed set of possible next states $\mathcal{Z}_{k+1} \subseteq \mathcal{Z}_p$ of $\mathcal{N}_p$ and sends it out after one step as the set of actual state $\mathcal{Z}_k \subseteq \mathcal{Z}_p$.

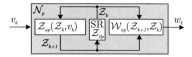

Figure 2.6.: Internal structure of a nominal nondeterministic I/O automaton

2.2.2. Output and transition degrees

$od(z)$ and $td(z', z)$, respectively, represent the output degree of z and the transition degree from state z to z'. $od(z)$ is the number of transitions leaving the state z [92]. $td(z', z)$ is the number of transitions connecting the state z to state z'. These concepts are formalized for I/O automata at a given state z as follows:

$$od(z) = \sum_{z'}^{\mathcal{Z}} \sum_{v_i}^{\mathcal{V}_a(z',z)} \sum_{w_i}^{\mathcal{W}_a(z',z)} L(z', w_i, z, v_i) \tag{2.35}$$

$$td(z', z) = |\{(v, w) \in \mathcal{V}_a(z', z) \times \mathcal{W}_a(z', z) : L(z', w, z, v) = 1\}| \tag{2.36}$$

$$= \sum_{v_i}^{\mathcal{V}_a(z', z)} \sum_{w_i}^{\mathcal{W}_a(z', z)} L(z', w_i, z, v_i). \tag{2.37}$$

2.2.3. Adjacency matrices

Since the powers of adjacency matrices give information about a walk or a state sequence of graph, as mentioned in [27, 92], they are adapted now to the I/O automata used in this thesis. This technique is similarly applied in Max-Plus algebra [38] and the reference [137] in a boolean and a symbolic fashion.

Numerical adjacency matrices

The weighted adjacency matrix of an automaton $\mathcal{N}$ is defined as a $(|\mathcal{Z}| \times |\mathcal{Z}|)$ matrix, written $(0, td)$-matrix,

$$\boldsymbol{A} = (a_{ij}) \text{ with } 1 \leq i, j \leq |\mathcal{Z}|, \tag{2.38}$$

where each entry a_{ij} equals the transition degree from state j to state i w.r.t. (2.37):

$$A = (a_{ij}) \text{ with } a_{ij} = td(i,j). \qquad (2.39)$$

Symbolic adjacency matrices

Consider the rings $(\mathbb{A}_z, \cup, \times)$, $(\mathbb{A}_v, \cup, \times)$ and $(\mathbb{A}_w, \cup, \times)$ with the three weighted adjacency matrices $A_z \in \mathbb{A}_z, A_v \in \mathbb{A}_v, A_w \in \mathbb{A}_w$ of an I/O automaton $\mathcal{N}$. They are respectively defined as $(|\mathcal{Z}| \times |\mathcal{Z}|)$ matrices as follows:

- $A_z = (a_{ij})$ with $a_{ij} = j \cap \mathcal{Z}_a(i)$.

- $A_v = (a_{ij})$ with $a_{ij} = \mathcal{V}_a(j,i)$.

- $A_w = (a_{ij})$ with $a_{ij} = \mathcal{W}_a(j,i)$.

Note that the interpretation of symbolic adjacency matrix entries of A_z, A_v and A_w is transposed compared to the numerical adjacency matrix A. This transposition is necessary to obtain correctly ordered sequences of states, inputs and outputs in A_z^k, A_v^k and A_w^k with $k > 1$.

Since the elements a_{ij} of the matrices above are symbols, the following subrings $(\mathcal{Z}, \cup, \times)$, $(\mathcal{V}, \cup, \times)$ and $(\mathcal{W}, \cup, \times)$ of the rings defined above, are considered for the matrix multiplication. The $\times$ represents the Cartesian product operator which is equivalent to a noncommutative concatenation of symbols often represented by the $\cdot$ symbol. That is for two given matrix entries a and b, $ab = a \cdot b = a \times b = \{a\} \times \{b\} = \{(a,b)\}$. The $\cup$ operator is the commutative union operator. For two given matrix entries a and b, $a \cup b = \{a\} \cup \{b\} = \{a,b\}$. Hence, the entries a_{ij} of the products A_z^k, A_v^k and A_w^k consists of sets of states, inputs and output sequences of the length k respectively from state i to state j with $i, j = 1 \ldots |\mathcal{Z}|$. The product of matrices is applied according to the following rule:

$$A_1 A_2 = \left(\bigcup_{\nu=1}^{|\mathcal{Z}|} a_{1,\mu\nu} \times a_{2,\nu\lambda} \right) = (a_{3,\mu\lambda}) = A_3 \text{ with } \mu, \lambda = 1 \ldots |\mathcal{Z}|. \qquad (2.40)$$

This rule is similar to the well-known matrix multiplication where the Cartesian product $\times$ acts as a noncommutative multiplication and the $\cup$ operator as an addition.

Example. Consider the I/O automaton of Fig. 2.7. It has a transition degree of one between every states except from state 3 to 4. Figure 2.8 shows the corresponding graph with transition degrees as labels. For this example, the numerical adjacency matrix matrices are

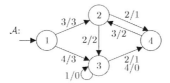

Figure 2.7.: Example of an I/O automaton for adjacency matrices

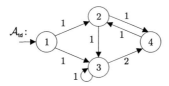

Figure 2.8.: Example of a transition degree graph

given in (2.41).

$$A = \begin{pmatrix} 0 & 0 & 0 & 0 \\ 1 & 0 & 0 & 1 \\ 1 & 1 & 1 & 0 \\ 0 & 1 & 2 & 0 \end{pmatrix}, A^2 = \begin{pmatrix} 0 & 0 & 0 & 0 \\ 0 & 1 & 2 & 0 \\ 2 & 1 & 1 & 1 \\ 3 & 2 & 2 & 1 \end{pmatrix}, A^3 = \begin{pmatrix} 0 & 0 & 0 & 0 \\ 3 & 2 & 2 & 1 \\ 2 & 2 & 3 & 1 \\ 4 & 3 & 4 & 2 \end{pmatrix}. \quad (2.41)$$

The matrix entry $A^3(2,1) = 3$ shows that there are three possible paths from state 1 to state 2 within 3 steps.

Graphs of the symbolic adjacency matrices for states, inputs and outputs are depicted in Fig. 2.9, Fig. 2.10, and Fig. 2.10, respectively. The symbolic state adjacency matrices are

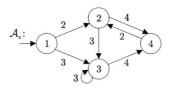

Figure 2.9.: Graph of the symbolic state matrix of Fig. 2.7

given in (2.42).

$$A_z = \begin{pmatrix} \emptyset & 2 & 3 & \emptyset \\ \emptyset & \emptyset & 3 & 4 \\ \emptyset & \emptyset & 3 & 4 \\ \emptyset & 2 & \emptyset & \emptyset \end{pmatrix}, A_z^3 = \begin{pmatrix} \emptyset & (24 \cup 34)\,2 & (3^2 \cup 23)\,3 & (3^2 \cup 23)\,4 \\ \emptyset & 342 & 3^3 \cup 423 & 3^2\,4 \cup 424 \\ \emptyset & 342 & 3^3 \cup 423 & 3^2\,4 \cup 424 \\ \emptyset & 242 & 23^2 & 234 \end{pmatrix}.$$

(2.42)

The matrix entry $A_z^3(1,2) = \{(24 \cup 34)2\} = \{(2,4,2);(3,4,2)\}$ contains possible state sequences from state 1 to state 2 within 3 steps.

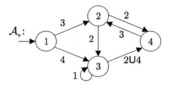

Figure 2.10.: Graph of the symbolic input matrix of Fig. 2.7

The symbolic input adjacency matrices are given in (2.43) and (2.44).

$$A_v = \begin{pmatrix} \emptyset & 3 & 4 & \emptyset \\ \emptyset & \emptyset & 2 & 2 \\ \emptyset & \emptyset & 1 & 2 \cup 4 \\ \emptyset & 3 & \emptyset & \emptyset \end{pmatrix},$$

(2.43)

$$A_v^3 = \begin{pmatrix} \emptyset & (4\,(2 \cup 4) \cup 32)\,3 & (41 \cup 32)\,1 & (41 \cup 32)\,(2 \cup 4) \\ \emptyset & 2\,(2 \cup 4)\,3 & 21^2 \cup 232 & 32^2 \cup 21\,(2 \cup 4) \\ \emptyset & 1\,(2 \cup 4)\,3 & 1^3 \cup (2 \cup 4)\,32 & 1^2\,(2 \cup 4) \cup (2 \cup 4)\,32 \\ \emptyset & 323 & 321 & 32\,(2 \cup 4) \end{pmatrix}.$$

(2.44)

The matrix entry $A_v^3(1,2) = \{(4(2 \cup 4) \cup 32)3\} = \{(4,2,3);(4,4,3);(3,2,3)\}$ contains possible input sequences from state 1 to state 2 within 3 steps.

The symbolic output adjacency matrices are given in (2.45) and (2.46) as follows:

$$A_w = \begin{pmatrix} \emptyset & 3 & 3 & \emptyset \\ \emptyset & \emptyset & 2 & 1 \\ \emptyset & \emptyset & \emptyset & 1 \cup 0 \\ \emptyset & 2 & \emptyset & \emptyset \end{pmatrix},$$

(2.45)

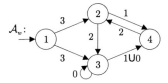

Figure 2.11.: Graph of the symbolic output matrix of Fig. 2.7

$$A_w^3 = \begin{pmatrix} \emptyset & (3\,(0 \cup 1) \cup 13)\,2 & (0\,3 \cup 2\,3)\,0 & (0\,3 \cup 2\,3)\,(0 \cup 1) \\ \emptyset & 2\,(0 \cup 1)\,2 & 0^2\,2 \cup 1\,2^2 & 1^2\,2 \cup 0\,2\,(0 \cup 1) \\ \emptyset & 0\,(0 \cup 1)\,2 & (0 \cup 1)\,2^2 \cup 0^3 & 0^2\,(0 \cup 1) \cup (0 \cup 1)\,1\,2 \\ \emptyset & 1\,2^2 & 0\,2^2 & 2^2\,(0 \cup 1) \end{pmatrix}. \quad (2.46)$$

The matrix entry $A_w^3(1,2) = \{(3(0 \cup 1) \cup 13)2\} = \{(3,0,2);(3,1,2);(1,3,2)\}$ contains possible output sequences from state 1 to state 2 within 3 steps.

2.2.4. W-determinism

The *output generation* of an automaton $\mathcal{N}$ is said to be *deterministic* when the output w is unique for a given state-input combination (z,v). An I/O automaton $\mathcal{N}$ with a deterministic output generation is said to be W-deterministic.

Lemma 2.1. *A nondeterministic I/O automaton $\mathcal{N}$ is W-deterministic iff*

$$\forall\,(z,v) \in \mathcal{Z} \times \mathcal{V}, \mathcal{Z}_a(z,v) \neq \emptyset \Rightarrow |\mathcal{W}_a(z,v)| = 1. \quad (2.47)$$

Proof. See Appendix B, Page 207. □

2.2.5. I/O trellis automata

Let Fig. 2.12 describe an unfolding of the I/O automaton from Fig. 2.1 over the time steps k for the set of state sequences

$$\mathcal{Z}(0 \ldots 4) = \{Z_1, Z_2, Z_3\} = \{(1,2,1,2,4),(3,1,2,4,3),(3,1,2,4,2)\} \quad (2.48)$$

This automata unfolding, subsequently called I/O trellis automata, adequately reflects the possible state trajectories of the automaton $\mathcal{A}_p$ of Fig. 2.1, redundant paths as well as their length.

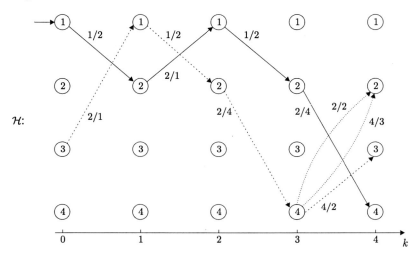

Figure 2.12.: Example of the I/O trellis automaton resulting from unfolding $\mathcal{A}_p$ of Fig. 2.1 w.r.t. (2.48)

In order to formally show the relationship between classic I/O automata and I/O trellis automata, the latter is formalized in a similar fashion as the former. This permits to formalize the unfolding procedure in the sequel.

A trellis basically consists of a succession of sets of states $\mathcal{Z}_k$ for each step k as in Fig. 2.12. Compared to nondeterministic automata of Section 2.1, it shows along which states the process may have gone through until a given step k. Hence, trellis automata have a kind of memory.

Definition 2.6. *The I/O trellis automaton is defined as*

$$\mathcal{H} = (\mathcal{Z}, \mathcal{V}, \mathcal{W}, \eta, \mathcal{Z}(0 \cdots \boldsymbol{K_e})). \tag{2.49}$$

η represents the characteristic function:

$$\eta : \mathcal{Z} \times \mathcal{W} \times \mathcal{Z} \times \mathcal{V} \times \mathbb{N} \rightarrow \{0, 1\}$$

$$\eta(z', w, z, v, k) = \begin{cases} 1, & (z', w, z, v, k)! \\ 0, & else. \end{cases} \tag{2.50}$$

Unfolding formalization. Now, the automaton unfolding formula to derive an I/O trellis graph $\mathcal{H}$ from an I/O automaton $\mathcal{N}$ w.r.t. a set of state sequences (2.5) is described as follows:

$$\eta(z', w, z, v, k) = L(z', w, z, v) \cdot \left[\bigvee_{Z_i(0\ldots k_{ei})}^{\mathcal{Z}(0\cdots\boldsymbol{K_e})} ((Z_i(k) \equiv z) \wedge \right. \\ \left. (Z_i(k+1) \equiv z')) \right] \tag{2.51}$$

with $w \in \mathcal{W}_a(z', z), v \in \mathcal{V}_a(z', z)$.

Note that $\eta(z', w, z, v, k) = 1$ only if

1. $L(z', w, z, v) = 1$, i.e., the transition (z', w, z, v) is modeled in $\mathcal{N}_p$, and

2. both predicates $(Z_i(k) \equiv z)$ and $(Z_i(k+1) \equiv z')$ are TRUE for at least one state sequence $Z_i(0\cdots k_{ei}) \in \mathcal{Z}(0\cdots\boldsymbol{K_e})$.

The unfolding procedure obviously leads to a graph which better visualizes the dynamic behavior of a system or its redundancies compared to the classic graph of an I/O automaton. However, classic graphs of I/O automata have no redundant transitions as trellis graphs, thus, they are less complex and better suited for computation than trellis automata.

Particular properties of I/O trellis. Trellis automata used in this thesis have the following properties which slightly differ from the literature as follows:

1. Every state of the trellis that is not a final state has a finite output degree.

$$\forall z \notin \mathcal{Z}_F, 0 \leq od(z) < \infty. \tag{2.52}$$

2. The maximal number of transitions between two states is finite.

$$\forall (z', z) \in \mathcal{Z} \times \mathcal{Z}, 0 < td(z', z) < \infty. \tag{2.53}$$

3. All reflected state sequences must not have the same length.

$$Z_i \neq Z_j \Rightarrow k_{ei} \geq k_{ej} \vee k_{ei} < k_{ej}, \forall (Z_i, Z_j) \in \mathcal{Z}(0\cdots\boldsymbol{K_e})^2. \tag{2.54}$$

2.3. Discrete-event modeling issues

This sections discusses some issues often encountered during the modeling process of discrete-event systems:

- Distinction between controllable and observable events: How should they be interpreted for I/O automata?

- State space explosion: How should a discrete-event fault-tolerant control method deal with system complexity?

2.3.1. Controllable and observable events

This section highlights the difference between the event classification known from standard automata and I/O automata. In the models used in this thesis, there is no classification among controllable and observable events as it is the case in several articles from literature.

Particularly, the fact that an event is not controllable in the standard automata theory is expressed through nondeterministic I/O transitions in I/O automata, where the next state and/or the output can not be predicted in advance. Thus, such transitions are not controllable in the sense of standard automata. However, they are always observable since they are explicitly modeled as output of the plant. On the other side, control input events of an I/O automaton are always controllable in the sense of standard automata. Indeed, the controller of a plant can clearly decide to send a command w_c or not to do so.

The aforementioned properties of input and output events of I/O automata are crucial to guarantee the exact diagnosis required by the fault-tolerant control scheme of Fig. 1.1.

2.3.2. State space explosion

A common problem to discrete-event systems is the state space explosion problem. A modularisation of the problem through a splitting of the plant into several subsystems which are then composed, reduces the modeling efforts but not the overall complexity which is still nonpolynomial due to the cartesian products of the states.

A second problem is the ability or the possibility to find "good" models for the task at hand. A "good" model is an abstraction of a physical system without inessential details [118].

To deal with the state space explosion, heuristic physical properties of the plant are considered while modeling. Since the fault-tolerant control is the main objective, only those states, inputs and outputs which are relevant for that purpose are included in the model. The goal is usually to keep the model as exact as possible but as complex as necessary. This modeling philosophy often leads the system modeler to exclude redundant states, inputs and outputs. In fault-tolerant control scheme, those redundancies are necessary although they might not always be sufficient. Therefore, a suitable modeling philosophy for fault-tolerant systems must include redundancies, if available, despite the costs of complexity.

The on-line control reconfiguration method developed in this document deals with the complexity of systems by reducing the reconfiguration problem to the level of critical transitions, instead of the whole graph.

3. Discrete-event fault modeling

Abstract. *The aim of this chapter is to present a method for representing faults in a technological process. The method permits the computation of a model of the faulty plant based on the model of the nominal plant and the knowledge on the fault. The local effect of an actuator fault, a sensor fault or a system internal fault is used to compute the global behavior of the faulty plant.*

3.1. Faulty behavior and diagnosis approach

The main idea is summarized by the quote "Matter is neither created nor destroyed but transformed" [116]. This statement from Lavoisier reflects in a broader sense the behavior of a faulty plant and is very close to the modeling technique presented here. In fact, when a fault occurs, the intuitive interpretation of the consequences on the model would be the disappearing of a transition in $\mathcal{N}_p$. However, this is not quite correct because the signal sent by or to the controller does not just disappear but is transformed into a value that leads to an inconsistent, an unexpected or a forbidden behavior of the plant and/or the controller, or to a blocking control loop.

The diagnosis concept widely established in literature for quantitative and qualitative systems is depicted in Fig. 3.1. This chapter is dedicated to qualitative model-based diagnosis techniques. Analytical models represented by nondeterministic I/O automata are used. One established diagnosis technique relies on a consistency-based check making use of the model of the nominal behavior of the plant $\mathcal{N}_p$ and a model of the faulty behavior $\mathcal{N}_p^f$ for every possible fault $f \in \mathcal{F}$ (Fig. 3.1). The model of the fault-free behavior $\mathcal{N}_p$ is sufficient for fault detection which only answers the question whether a fault has occurred or not (see [49, 62, 162] and [171]).

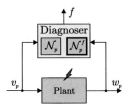

Figure 3.1.: Diagnosis of a discrete-event process

3.2. Motivation

A fault is detected whenever the measurements of the inputs v_p and outputs w_p are inconsistent with the nominal behavior of the plant $\mathcal{N}_p$. That is, the plant deviates from the required behavior. However, this inconsistency with $\mathcal{N}_p$ is necessary but not sufficient to identify the fault that occurred. For a given set of faults $\mathcal{F} = \{f_1, f_2\}$ a consistency between the measurements and $\mathcal{N}_p^{f_1}$ makes f_1 a fault candidate. Only an inconsistency between the measurements, $\mathcal{N}_p$ and $\mathcal{N}_p^{f_2}$ confirms the presence of a fault and excludes f_2 from the list of fault candidates. Thus, isolates the fault f_1. Several works, as [129, 169], have used this fault isolation concept for unobservable fault events and additional fault parameters in the transition functions respectively. This fault identification scheme shows the necessity of models of the faulty behavior.

In addition, the Fault Propagation Analysis (FPA) also make use of a model of the faulty behavior to study the roots and consequences of a fault on a system. The verdict can be used to discover weaknesses of the plant leading to severe consequences and achieve preventive improvements. Furthermore, model-based reconfiguration techniques always make use of a model of the faulty plant to determine the appropriate actions to take in order to fulfill a specification. This shows the importance of a model of the faulty system for different analysis tasks. [62] and [60] give surveys underlining the relevance of fault models for model-based diagnosis.

This chapter presents a method to obtain the model of such a faulty behavior for actuator, sensor and internal system faults in a systematic way. The quality of the model is an essential criterion for the success of a model-based diagnosis. Hence, building appropriate models of the faulty system is a major task of fault diagnosis. This is the main concern of this chapter. [151, 167] stressed the difficulty of building the right model due to the changing context, task and need of reusability which are in general conflicting goals.

However, the model of the faulty behavior $\mathcal{N}_p^f$ presented here is suitable for on-line and off-line diagnosis purposes.

A common problem is the fact that each fault leads to a separate model and a possibly large amount of data. In addition, it is arduous to heuristically derive a faulty model out of a nominal monolithic model which is usually obtained by composition. A fault in a single component requires an extensive investigation of the consequences on the global plant by considering the interactions among the components, the composition rule and the expertise of the modeler.

This chapter proposes to reduce the load of this modeling task by means of error-relations describing actuator, sensor, and system internal faults or failure. In general, the fault modeling task requires a deep knowledge of the system at hand which can also consist of subsystems. Hence, also the interaction among the subsystems has to be mastered in order to predict the faulty behavior of the overall plant. There is no need to stress how difficult it becomes with growing complexity. This motivates the development of the systematic fault modeling method presented next.

3.3. Fault definition and classification

3.3.1. Definitions

The definitions used here have been selected from different works in the literature, see [49, 112, 169] and [172]. A *fault* is an unpermitted deviation of at least one characteristic property or parameter of the system from the acceptable, usual or standard condition. A *failure* is a permanent interruption of a system's ability to perform a required function. A *symptom* is a change of an observable quantity from the normal behavior. In a cause-effect chain, the fault is one or more of the potential causes whereas the symptoms represents the effects.

Usually the FDI-techniques found in literature do not model the real causes of the faults but the effects observed after the fault occurred. As in literature, symptoms are used here to describe faults on a model level. In order to ease the readability, faults or failures always refer to these symptoms in the sequel. Recall that faults considered in this thesis are supposed to be identifiable, thus, the diagnosability condition is assumed to be always satisfied.

3.3.2. Fault classification

The plant is decomposed in three main parts which can be subject of faults:

- actuators which control the process: a faulty actuator has an altered actuator behavior, that is an altered input to the process.

- the process: a faulty process with nominal actuators and sensors means a nominal input may drive the process into a wrong state $z'_\sharp$. This is a faulty destination despite the correct I/O measurements.

- sensors: a faulty sensor falsifies the output of the process, hence, represents an alteration of the output labels.

Henceforth, it is reasonable to define a fault classification with respect to the impact they have on the model as follows:

- Faulty I/O labels: they consist of faults *altering the input/ouput behavior* v/w into $v_\sharp/w$, $v/w_\sharp$ or $v_\sharp/w_\sharp$ where v and w represent the nominal signals while $v_\sharp$ and $w_\sharp$ are for the faulty ones from a state z. The next state z' which is the destination of the transition remains unchanged.

- Faulty destinations: they change a transition from a state z to the faulty next state $z'_\sharp$ instead of the nominal one z'. The I/O labels remain unchanged.

- Appearing (disappearing) transitions in (from) the nominal model $\mathcal{N}_p$.

- Faulty *behaviors*: they are represented by output symbols which are consistent with the behavior of $\mathcal{N}_p^f$ whereas they are inconsistent with the nominal behavior of $\mathcal{N}_p$.

This thesis proposes a method which incorporates faulty input/output labels, faulty destinations, appearing and disappearing transitions into one faulty behavior represented by $\mathcal{N}_p^f$.

3.4. Case study: a fluid level control process subject to faults

Figure 3.2 illustrates the fluid level control of Section 1.3.1 extended with two kinds of fault also represented by the bolt symbol: namely the defect sensor of level 4 and the leakage in the tank T_1. Each situation respectively describes an input, output and a state fault.

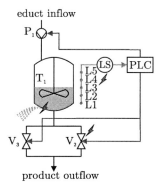

Figure 3.2.: Mixture preparation process

First, the nominal model of the plant $\mathcal{N}_p$ depicted in Fig. 3.3 is described. The pump P_1 and the valves V_2, and V_3 are the relevant actuators of the system controlled by the signals v_p. The input $v_p = 0$ models the command "close all valves" whereas $v_p = i$ is the command "open valve V_i and close the others". The states z_p of the plant $\mathcal{N}_p$ are modeled by the states of the tank T_1 which vary from 0 (empty) up to 5 (full). Five level sensors (LS) permit a discrete measurement modeled by the output w_p of the level of the tank T_1. Let w_p model the result of an inflow or an outflow of the educt, so that $w_p = z'_p$ holds.

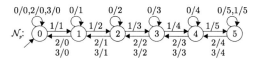

Figure 3.3.: Complete I/O automaton graph of the fluid level control process

Recall that the control objective of the process is to fill the tank T_1 from level 0 up to level 4, then down to level 1 and back to level 4 in a cyclic way.

A heuristic analysis of the faults leads to the following conclusions:

- The blocking of V_2 causes a stagnation of the educt at the level where this fault occurs.

- The defect sensor of level 4 makes that level unobservable, hence, the controller will keep the pump P_1 open during the filling sequence. If this happens during an emptying sequence, the sensor fault will take effect by the next filling sequence.

- Assume that the leakage in tank T_1 reduces the level of educt of one notch at each time step k. Consequently, during a filling sequence the educt inflow will compensate the outflow caused by the leakage, resulting into a stagnation. During an emptying sequence, the level of educt will sink faster than expected and remain at level 1 where the controller would try to fill up the tank again without success.

Now, an approach to model the effect of this kind of faults in a systematic way by modeling the faulty behavior of actuators, sensors and the process in order to obtain models of the overall faulty plant is proposed.

3.5. Modeling of faults and failures

3.5.1. Main idea

The fault modeling concept presented in this chapter consists of finding a way to determine how the nominal plant would behave under the presence of actuator, sensor and system internal faults. For an actuator fault the question to be answered is how would the nominal plant behave if its input was corrupted by a fault like a toggled bit, a bad communication channel or a cable break which are common faults in industrial environments. The *input-error relation* Err_v is introduced to answer this question. It exploits the fact that the faulty plant $\mathcal{N}_p^f$ behaves similarly to the nominal plant $\mathcal{N}_p$ receiving a corrupted input. The same idea is extended to sensor faults for an introduced *output-error relation* Err_w which represents a falsification of the nominal sensor value w_p sent to the controller. Concerning system internal faults, the *state-error relation* is introduced to model faulty state destinations.

3.5.2. Actuator faults and failures

The input-error relation Err_v is introduced now to model the alternation of the nominal input $v_p = w_c$ sent to the plant by the controller:

$$Err_v \subseteq \mathcal{V}_p \times (\mathcal{V}_p \cup \{\varepsilon\}). \tag{3.1}$$

The input-error operator $E_v(\cdot)$ is defined in (3.2) to describe faulty input events v_f which are in a relation with the nominal input event v_p through Err_v. If an actuator is considered as faulty, then the input-error operator transforms the nominal value v_p into a corrupted one

$v_\sharp \in \mathcal{V}_p$. In case of an input failure, the input signal is supposed to be out of the reserved range or unknown. Thus, it is supposed to equal the empty symbol ε:

$$
\begin{aligned}
E_v(v_p) &= v_f \text{ with } v_f \in \mathcal{V}_p \cup \{\varepsilon\} : (v_p, v_f) \in Err_v \\
\text{and } v_f &= \begin{cases} v_p = w_c & \text{faultless case} \\ v_\sharp \neq w_c & \text{faulty case} \\ \varepsilon \neq w_c & \text{input failure.} \end{cases}
\end{aligned}
\tag{3.2}
$$

3.5.3. Sensor faults and failures

The output-error relation Err_w models the alteration of the nominal output $w_p = v_c$ sent to the controller:

$$
Err_w \subseteq \mathcal{W}_p \times (\mathcal{W}_p \cup \{\varepsilon\}).
\tag{3.3}
$$

The output-error operator $E_w(\cdot)$ is defined in (3.4) similarly to the input-error operator:

$$
\begin{aligned}
E_w(w_p) &= w_f \text{ with } w_f \in \mathcal{W}_p \cup \{\varepsilon\} : (w_p, w_f) \in Err_w \\
\text{and } w_f &= \begin{cases} w_p & \text{faultless case} \\ w_\sharp \neq w_p & \text{faulty case} \\ \varepsilon \neq w_p & \text{output failure.} \end{cases}
\end{aligned}
\tag{3.4}
$$

3.5.4. System internal faults and failures

A system internal fault occurs when neither the actuator nor the sensor but the process is subject to a fault, e.g., a leakage in a tank. The state-error relation $Err_{z'}$ is introduced to model faulty state transition w.r.t. a specific input.

$$
Err_{z'} \subseteq \mathcal{Z}_p \times \mathcal{V}_p \times (\mathcal{Z}_p \cup \{z_\varepsilon\})
\tag{3.5}
$$

The state-error operator $E_{z'}(\cdot)$ is defined in (3.6) so that it only influences the next state transition in the plant automaton. This results in an alternation, an appearance or a deletion of states transition:

$$
\begin{aligned}
E_{z'}(z_p, v_p) &= z'_f \text{ with } z'_f \in \mathcal{Z}_p \cup \{z_\varepsilon\} : (z_p, v_p, z'_f) \in Err_{z'} \\
\text{and } z'_f &= \begin{cases} \mathcal{Z}_{ap}(z_p, v_p) & \text{faultless next state} \\ z'_\sharp \notin \mathcal{Z}_{ap}(z_p, v_p) & \text{faulty next state} \\ z'_\varepsilon \notin \mathcal{Z}_p & \text{next state failure.} \end{cases}
\end{aligned}
\tag{3.6}
$$

3.6. Construction of the model of a system subject to faults and failures

This section explains how the error maps $E_v(\cdot)$, $E_w(\cdot)$ and $E_{z'}(\cdot)$ are used to derive the characteristic function of the faulty behavior L_p^f which reveals which transitions belong to $\mathcal{N}_p^f$ and which not.

3.6.1. Input failure

It can be assumed without loss of generality that an input failure leads to a self-loop at the state z_p where the failure occurred and a generation of the possible outputs $\mathcal{W}_{ap}(z_p, z_p)$ which are consistent with the self-loop. The generation of $\mathcal{W}_{ap}(z_p, z_p)$ means that the sensors still work properly and would eventually reveal the presence of an input failure after running a diagnosis algorithm. In the case that there exists no self-loop or the measured value can not be recognized by the sensors, the output is ε. Thus, faulty transitions are obtained with

$$
\begin{aligned}
&L_p^f(z_p, w_p, z_p, v_p) = (E_v(v_p) \equiv \varepsilon) \wedge (\mathcal{V}_{ap}(z_p, z_p) \neq \emptyset), \\
&\forall (z_p, w_p) \in \mathcal{Z}_p \times \{\mathcal{W}_{ap}(z_p, z_p)\}.
\end{aligned}
\tag{3.7}
$$

With (3.7), the characteristic function of the faulty plant L_p^f equals one only if both predicates are TRUE otherwise it equals zero. Equation (3.7) represents the case where a transition (z_p', w_p, z_p, v_p) disappears from $\mathcal{N}_p$ and emerges as a self-loop (z_p, w_p^f, z_p, v_p) in $\mathcal{N}_p^f$ with $w_p^f \in \mathcal{W}_{ap}(z_p, z_p)$. If $\mathcal{W}_{ap}(z_p, z_p) = \emptyset$, the empty symbol is used, i.e, $w_p^f = \varepsilon$.

3.6.2. Output failure

If the plant is controlled in a closed-loop, an output failure would affect the process after a finite number of steps because the feedback controller would not get the expected value but the symbol $\varepsilon = E_w(w_p)$. In this case the behavior of the control loop after an output failure can not be predicted because it depends on the control policy implemented in the feedback controller for the output ε from the plant.

In the case of an open-loop, an output failure would have no impact on the process since the controller would not be aware of it due to the missing feedback connection. The characteristic function of faulty transitions for output failures reflects a replacement of every transition labeled with the failed output w_p by ε as follows:

$$L_p^f(z_p', \varepsilon, z_p, v_p) = L_p(z_p', w_p, z_p, v_p) \wedge (E_w(w_p) \equiv \varepsilon),\ \forall(z_p', z_p, v_p) \in \mathcal{Z}_p^2 \times \mathcal{V}_p. \quad (3.8)$$

Equation (3.8) represents the case where a transition (z_p', w_p, z_p, v_p) from $\mathcal{N}_p$ is replaced in $\mathcal{N}_p^f$ by $(z_p', \varepsilon, z_p, v_p)$.

3.6.3. State failure

This situation occurs when the process reaches a state z_ε' which is not covered by the model $\mathcal{N}_p$, although the input v_p is correct. Consequently, the model $\mathcal{N}_p$ remains at the same state z_p since z_ε' is unknown. The nominal sensors correspondingly send a measurement w_i in accordance with z_ε', which might belong to $\mathcal{W}_p$ or not. Thus, $w_i \in \mathcal{W}_p \cup \{\varepsilon\}$ after a state failure. This is expressed by

$$L_p^f(z_\varepsilon', w_p, z_p, v_p) = (E_{z'}(z_p, v_p) \equiv z_\varepsilon'),\ w_p \in \mathcal{W}_p \cup \{\varepsilon\}. \quad (3.9)$$

3.6.4. Computation of the faulty behavior

This section presents the computation of $\mathcal{N}_p^f$ by means of the error maps previously introduced for a plant subject to faults instead of failures discussed in Sections 3.6.1, 3.6.2 and 3.6.3. The resulting internal view of the faulty nondeterministic automaton is given in Fig. 3.4. It shows that the input-error map $E_v(\cdot)$ has a direct impact on the new active state function $\mathcal{Z}_{ap}^f(\cdot)$ and an indirect impact on the new faulty output function $\mathcal{W}_{ap}^f(\cdot)$. It also shows that the output-error map $E_w(\cdot)$ influences the new active output function $\mathcal{W}_{ap}^f(\cdot)$.

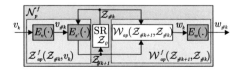

Figure 3.4.: Internal structure of a faulty nondeterminisitic I/O automaton

There are two main aspects to consider in order to determine the faulty next state set $\mathcal{Z}_{ap}^f(\mathcal{Z}_{\sharp k}, v_k)$ of Fig. 3.4, namely:

- the influence of a faulty actuator signal $v_{\sharp i}$ on the nominal state transition formalized by $\mathcal{Z}_{ap}(\mathcal{Z}_{\sharp k}, v_{\sharp i})$ w.r.t. the input-error map,

- the influence of a faulty actuator signal $v_{\sharp i}$ on a faulty state transition formalized by the successive evaluation of $E_v(\cdot)$ and $E_{z'}(\cdot)$. This can be used to model the behavior of a plant which is simultaneously subject of an actuator and a system internal fault. Thus, this approach is not limited to faults which do not occur together.

The next state set of the faulty system model belongs to the set

$$\mathcal{Z}_{ap}^f(\mathcal{Z}_{\sharp k}, v_k) = \bigcup_{z_k}^{\mathcal{Z}_k} E_{z'}(z_k, E_v(v_k)). \tag{3.10}$$

The outputs of the faulty system are obtained by applying the output-error operator $E_w(\cdot)$ on every output in the transitions contained in the pair $(\mathcal{Z}_{\sharp k+1}, \mathcal{Z}_{\sharp k})$. Hence, the faulty active output set is defined by

$$\mathcal{W}_{ap}^f(\mathcal{Z}_{\sharp k+1}, \mathcal{Z}_{\sharp k}) = \bigcup_{w_i}^{\mathcal{W}_{ap}(\mathcal{Z}_{\sharp k+1}, \mathcal{Z}_{\sharp k})} E_w(w_i) \text{ with } \mathcal{Z}_{\sharp k+1} \subseteq \mathcal{Z}_{ap}^f(\mathcal{Z}_{\sharp k}, v_k). \tag{3.11}$$

$\mathcal{Z}_{ap}^f(\cdot)$ and $\mathcal{W}_{ap}^f(\cdot)$ are now used in Equation (3.12) w.r.t (3.10) and (3.11) to determine the characteristic function L_p^f of the faulty system model $\mathcal{N}_p^f$ of Fig. 3.4 for a given transition (z', w, z, v) as follows:

$$\boxed{\begin{aligned} &\forall (z', w, z, v) \in \mathcal{Z}_p \times \mathcal{W}_p \times \mathcal{Z}_p \times \mathcal{V}_p, \\ &L_p^f(z', w, z, v) = \begin{cases} 1 & \text{if } z' \in \mathcal{Z}_{ap}^f(z, v) \wedge w \in \mathcal{W}_{ap}^f(z', z) \\ 0 & \text{else.} \end{cases} \end{aligned}} \tag{3.12}$$

3.7. Application on a fluid level control process subject to faults

Each of the following faulty cases is used to demonstrate how to build the characteristic function L_p^f depending on the considered fault w.r.t. (3.12).

3.7.1. Blocking valve V_2

When the valve V_2 is blocking, any attempt of the controller to open it, fails. It means that the opening command $v_p = 2$ has the same effect as the closing command $v_t = 0$. The corresponding input error-relation is defined as $E_v(2) = 0$ as shown in Tab. 3.1.

Table 3.1.: Input-error map for the blocking valve V_2

v	$v_f = E_v(v)$
0	0
1	1
2	0
3	3

Figure 3.3 and Equation (3.10) imply that the next state set of the state-input combination $(1, 2)$ is $\mathcal{Z}_{ap}^f(1, 2) = \bigcup_{z_k}^{\{1\}} E_{z'}(0, E_v(2)) = \mathcal{Z}_{ap}(1, 0) = \{1\}$. Thus, $(\mathcal{Z}_{k+1}, \mathcal{Z}_k) = (1, 1)$ holds in Fig. 3.4. According to (3.11) the faulty active outputs are $\mathcal{W}_{ap}^f(1, 1) = \bigcup_{w_i}^{\mathcal{W}_{ap}(1,1)} E_w(w_i) = E_w(1) = \{1\}$ w.r.t. (3.3) because there is no sensor fault. By applying (3.12), $L_p^f(1, 1, 1, 2) = 1$ and $L_p^f(0, 0, 1, 2) = 0$ are obtained. This formally shows how the original transition $1 \to 0$ for $v = 2$ is now transformed into the self-loop $1 \to 1$. From other states with $v = 2$, this procedures leads to similar self-loops within the dashed lines of Fig. 3.5.

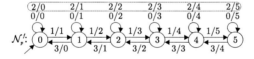

Figure 3.5.: Model of the fluid level control process with V_2 blocking

3.7.2. Sensor failure

This type of faults occurs when, e.g., a sensor is defective and no longer able to detect any physical change. This is modeled by the output error-relation $E_w(4) = \varepsilon$ as shown in Tab. 3.2.

In order to make use of (3.8) it is necessary to consider all the transitions where the output $w_p = 4$ is generated, i.e., $\forall (z_p', z_p, v_p) \in \mathcal{Z}_p^2 \times \mathcal{V}_p$ where $\{4\} \in \mathcal{W}_{ap}(z_p', z_p)$. Hence, (3.8) implies that $L_p^f(4, \varepsilon, 3, 1) = L_p^f(4, \varepsilon, 4, 0) = L_p^f(4, \varepsilon, 5, 2) = L_p^f(4, \varepsilon, 5, 3) = 1$ as depicted in Fig. 3.6.

Table 3.2.: Output-error relation

w	$w_\sharp = E_w(w)$
0	0
1	1
2	2
3	3
4	ε

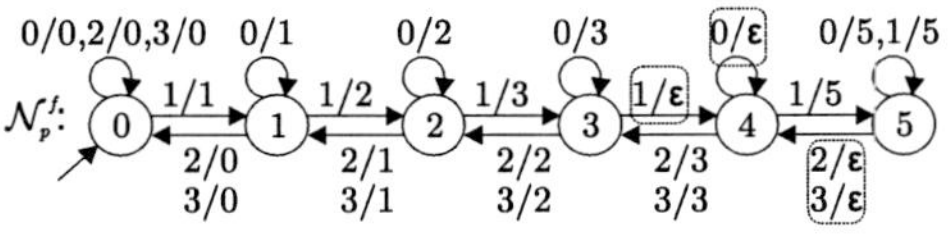

Figure 3.6.: Model of the fluid level control process with failure of sensor 4

3.7.3. Internal fault: leakage in tank T_1

Consider the leakage situation in tank T_1 as described in Section 3.4. The stagnation of educt caused by this fault can be formalized as $E_{z'}(z_p, 1) = z_p$, $\forall z_p \in \mathcal{Z}_p$. Moreover, the leakage will lead to a loss of educt even though all valves are closed. Therefore, $v_p = 0$ which models the "close all valves" command is equivalent with an opening command for valve V_2 or V_3. This actuator fault is modeled by $E_v(0) = 2$ whereas all sensors are working properly. Assume that the outflow due to the leakage is the same as the inflow through pump P_1. It is obvious that the leakage leads to a stagnation of the fluid during a filling sequence whereas it accelerates the emptying sequence. Since the model used here does not take the time into account, only faulty transitions are listed in Tab. 3.3.

Table 3.3.: State-error and input error relation

(z, v)	$z'_f = E_{z'}(z, v)$
(0,1)	0
(1,1)	1
(2,1)	2
(3,1)	3
(4,1)	4
(5,1)	5

v	$v_f = E_v(v)$
0	2
1	1
2	2
3	3

The following explains how the parallel effect of two faults can be modeled. Equation (3.10) implies, e.g., that $\mathcal{Z}_{ap}^f(2, E_v(1)) = \bigcup_{z_k}^{\{2\}} E_{z'}(2, E_v(1)) = \bigcup_{z_k}^{\{2\}} E_{z'}(2, 1) =$

$E_{z'}(2,1) = \{2\}$. Now, the effect of the additional actuator fault is considered. Equation (3.10) leads to $\mathcal{Z}^f_{ap}(2, E_v(0)) = \bigcup^{\{2\}}_{z_k} E_{z'}(2, E_v(0)) = E_{z'}(2,2) = \{1\}$. For both previously considered faults, (3.11) does not contribute to the construction of $\mathcal{N}^f_p$ because the sensors are supposed to be fault-free. Thus, the second part of (3.12) consists of $\mathcal{W}^f_{ap}(2,2) = \{2\}$ for the faulty next state and $\mathcal{W}^f_{ap}(1,2) = \{1\}$ leading to the faulty transitions $L^f_p(2,2,2,1) = 1$ and $L^f_p(1,1,2,0) = 1$ whereas the transition $(3,3,2,1)$ and $(2,2,2,0)$ disappear from $\mathcal{N}_p$ (Fig. 3.3). The application of the described steps on other transitions of $\mathcal{N}_p$ leads to the automaton of Fig. 3.7 which clearly shows that no matter at which state the considered leakage occurs, the level of educt will either stagnate or decrease.

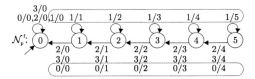

Figure 3.7.: Model of the fluid level control process with a leakage in tank T_1

3.8. Literature review

In continuous LTI-systems, a fault can be represented as an additional input to the system and be easily involved in the state space representation due to the superposition principle. Similar approaches have been used in discrete-event systems by extending the definition of automata for the nominal behavior with a new input representing the fault [112, 139]. A drawback of this method is the fact that every transition of the nominal automaton has to be studied to model the effect of the fault in a suitable way.

[114] expressed the signal falsification paradigm used here in terms of Byzantine faults which is nowadays a common expression among the computer scientists.

[169] obtained the model of the faulty behavior by means of a label propagation function comparable to the composition of the nominal standard automaton behavior with a fault automaton consisting of a nominal state and a faulty state. The transitions are chosen in such a way that after the composition only those states which can be reached by fault events are labeled with the F-symbol representing the fault. The concept of unobservable faulty events for diagnosis is investigated in [177] for timed I/O automata. Contrary to these approaches, faults are not modeled here by means of unobservable events since only events

which can be observed or registered by the sensors are considered in the following. Instead of a single unobservable event, a behavioral approach is used here. The fault is represented by a specific behavior of the plant which differs from the nominal one. Moreover, the approach presented here does not require an enlargement of the states of the nominal model with faulty states and new unobservable events, except for failures where the empty symbol ε is used. Only those events and states which are defined in the *complete* nominal model are reused in the faulty case with different transitions, e.g., new self-loops.

The approach proposed in [46] for Petri nets considers faults in components through explicit faulty places with associated labeled transitions. One goal was to find possible scenarios revealing which component could have been responsible for the observed fault. A similar idea is proposed in [142] for state charts. Instead of these explicit distinction of faulty states and transitions in advance, error-maps are used here and can be applied on every state or I/O transition according to the fault to be modeled.

Another approach consists of a sampling, abstraction and identification of the faulty dynamic of the plant under the influence of faults investigated in [129]. This is related to the data-driven approaches surveyed by [145]. This method has proven to provide the most complete behavior of the plant but is costly in terms of experiments, generation of fault prone test sequences and quantizer design which has to be performed. The diagnosability is often tested in literature via deterministic automata working as diagnosers and nondeterministic automata titled as verifiers. They consists of states labeled with Y or N revealing the occurrence of a fault or not, i.e, (z, Y) in the faulty case or (z, N) in the nominal case. The original construction procedure was given in [169] and an equivalent approach is proposed in [55, 56]. Contrary to the diagnoser or verifier from [169], faulty behavior automata presented here can not be used directly as diagnoser but can help to solve diagnosis problems.

Effects of a fault on the plant are modeled in [90, 137] by a set of deterministic models called family whereas the authors assume to have no information about the actually correct model.

4. Control design

Abstract. *A novel control design method for discrete-event systems described by I/O automata is proposed in this chapter. Based on the formalism of Chapter 2, a modeling method of the control loop is described. The order of events during an execution cycle of a control loop is explicitly discussed in order to explain the operational functioning of a control loop. Then, the property of the well-posedness of a control loop is addressed.*

The specification modeling and the feasibility of a specification are studied as crucial issues of the control design method. For feasible specifications, the control design procedure is explained and a realization scheme of the controller is presented.

4.1. Basic notions of the discrete-event control design

4.1.1. Problem statement

The control loop execution is based on the assumption of an unbuffered communication between the controller $\mathcal{N}_c$ and the plant $\mathcal{N}_c$ (Fig. 4.1). Adequate references to investigate the interaction between these two entities regarding synchronization, channel sharing and the use of buffers are [93, 94].

Figure 4.1.: Discrete-event control loop of I/O automata

Consider the discrete-event control loop of Fig. 4.1. Given is a plant modeled by an I/O automaton $\mathcal{N}_p$ and a safely feasible specification $\mathcal{S}$. The aim of the discrete-event control design (DECD) is to find a controller $\mathcal{N}_c$ with the following requirements:

1. Fulfillment of the specification $\mathcal{S}$ by the controlled plant.

2. Nonblockingness of the control loop despite the nondeterministic output function of the plant. This will be titled as the weak well-posedness of the control loop in Equation (4.10).

3. Determinism of the control output w_c at any step. This property is called W-determinism of the controller in Lemma 2.1, page 41.

This chapter proposes:

- a new design method of a discrete-event feedback controller $\mathcal{N}_c$

- an explicit realization scheme for the controller $\mathcal{N}_c$

- a controllability condition for the existence of $\mathcal{N}_c$.

Key issues regarding the feasibility of a specification and the determinism of the control output function were initially addressed in [1]. The former is extended in the sense that necessary and sufficient conditions are given for each specification type. The latter will be combined with the feasibility to express the controllability.

4.1.2. Model of the control loop

The control loop is modeled by a nondeterministic autonomous automaton $\mathcal{N}_l = \mathcal{N}_c/\mathcal{N}_p$. The building method of $\mathcal{N}_l$ is based on the assumption that $\mathcal{N}_c$ is W-deterministic. The control loop can then be defined as

$$\mathcal{N}_l = (\mathcal{Z}_l, \mathcal{W}, L_l, z_{0l}) \tag{4.1}$$

with

$$
\begin{aligned}
\mathcal{Z}_l &= \mathcal{Z}_c \times \mathcal{Z}_p, \text{ where } \boldsymbol{z}_l = (z_c, z_p)^T \text{ and } \boldsymbol{z'}_l = (z'_c, z'_p)^T \\
\mathcal{W}_l &= \mathcal{W}_p \\
z_{0l} &= (z_{0c}, z_{0p})^T \in \mathcal{Z}_l \\
L_l(\boldsymbol{z'}_l, w_p, \boldsymbol{z}_l) &= \bigvee_{v_p}^{v_p} L_c(z'_c, v_p, z_c, w_p) \cdot L_p(z'_p, w_p, z_p, v_p).
\end{aligned}
$$

4.1.3. Execution cycle of a control loop

In the following, the control loop is assumed to be constituted of two components in a feedback connection:

- A plant represented by a nondeterministic I/O automaton $\mathcal{N}_p$ and

- a controller represented by a W-deterministic I/O automaton $\mathcal{N}_c$.

It is important to clearly define how a *control loop cycle* must be understood in order to interpret the behavior of the control loop correctly. Also, the meaning of a *cycle* is relevant for practical purposes since the computing units implementing the control patterns usually execute them in a cyclic way. Thus, it is necessary to propose a cycle definition which is suitable not only for theoretical issues but also for practical implementation.

Assumed behavior and real behavior of a control loop. Figure 4.2 shows the order of signal evaluation in a control loop and opposes theoretical assumptions which are noncausal (see Fig. 4.2(a)) to the real behavior which is causal (see Fig. 4.2(b)).

So far, the input and output signals are assumed to occur simultaneously as shown in Fig. 4.2(a). Recall that this assumption concerns the behavior of an I/O automaton *alone*, i.e, not in a feedback connection like a control loop. Keeping this assumption in a control loop like Fig. 4.1 leads to an apparent noncausal behavior of the components. Thus, the real signal evaluation (Fig. 4.2(b)) of such a control loop is explained now between the two time steps k and $k + 1$.

At a step k, the plant and the controller are both in the states $Z_p(k)$ and $Z_c(k)$. In reality, the controller knows the target state into which the plant must be driven into in order to fulfill the specification. In the example of Fig. 4.2, the plant has to move from state $Z_p(k)$ to state $Z_p(k + 1)$. Based on the control law, the controller first sends the command $W_c(k)$ to the plant. The input signal of the plant V_p is then updated with $W_c(k)$ and subsequently triggers the state the state transition of the plant to the required state $Z_p(k + 1)$. Next, the output signal W_p is updated based on the latter state transition. This signal is then fed back to the controller as the input V_c. Finally, the controller state z_c is updated with the target state $Z_c(k + 1)$ which has now been reached. Depending on the next target state, the controller will generate the next command $W_c(k + 1)$. This procedure is then repeated for future values of k.

A detailed observation of the control loop behavior as previously explained shows that the dynamics of the plant $\mathcal{N}_p$ and the controller $\mathcal{N}_c$ in Fig. 4.1, during the modeling procedure, have to be interpreted differently. If the plant is modeled as a system which is in a state z_p and receives an input v_p, then the controller must be interpreted as a system which first send that command $w_c = v_p$ to the plant. The reaction of the plant as an output signal w_p, which is in accordance with its internal state transition, is then equal to the triggering controller input v_c for the internal state transition of the controller to z'_c. The latter influences the next decision of the controller. This is the behavior of the control loop, that will be used in this document. An example which exactly demonstrates the control behavior depicted in Fig. 4.2(b) is given in Chapter 7.

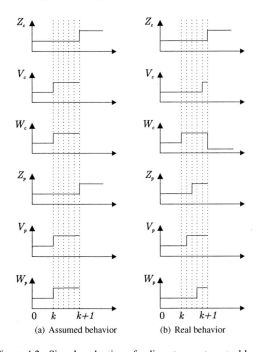

(a) Assumed behavior   (b) Real behavior

Figure 4.2.: Signal evaluation of a discrete event control loop

This section focussed on the the control loop behavior in the nominal case. Since the main goal of this thesis deals with fault tolerance, the behavior of the faulty control loop is also of interest. If a fault occurs, the plant would eventually generate output signal sequences W_p which would no longer be processed by the controller after a finite number of steps. This situation will be titled as *blocking* in the sequel. In Fig. 4.2(b), if the controller blocks e.g. at step k, it would maintain the last command $W_c(k-1)$ and no state transition or wrong state transition would be observed from k to $k+1$.

4.1.4. Control loop analysis

A nondeterministic I/O automaton $\mathcal{N}$ blocks when the characteristic function vanishes for some state-input combination (z, v). A control loop (Fig. 4.1) is said to be blocking whenever either the plant $\mathcal{N}_p$, the controller $\mathcal{N}_c$ or both block. This matter is handled by the following definition.

Definition 4.1 (Blocking automaton). *For an input sequence $V(0 \cdots k_e)$, a nondeterministic I/O automaton $\mathcal{N}$ is said to be blocking iff*

$$\bigwedge_{k=0}^{k_e} \bigvee_{z_{k+1}} \bigvee_{w_k}^{w} \bigvee_{z_k}^{z} L(z_{k+1}, w_k, z_k, V(k)) = 0. \tag{4.2}$$

If (4.2) does not hold, $\mathcal{N}$ is said to be *nonblocking* for the input sequence $V(0 \cdots k_e)$.

Blocking control loop and well-posedness. A control loop blocks in state $z_l(k)$ if $L_l(z_l(k+1), w_p(k), z_l(k)) = 0$ for all $z_l(k+1) \in \mathcal{Z}_l, w_p(k) \in \mathcal{W}_l$. This is the case when the plant $\mathcal{N}_p$ and/or the controller $\mathcal{N}_c$ blocks at a given state $z_c(k)$ or $z_p(k)$ for a given input $v_c(k)$ or $v_p(k)$ w.r.t Definition 4.1. In other words, there exists a state combination $(z_c(k), z_p(k))$ for which a given input combination $(v_c(k), v_p(k))$ leads to a blocking plant automaton $\mathcal{N}_p$, a blocking controller automaton $\mathcal{N}_c$ or both.

Well-posedness of a control loop. The question of the well-posedness of a control loop becomes relevant when considering a plant with a controller in a feedback connection for which

$$w_p = v_c \wedge w_c = v_p \tag{4.3}$$

holds. An "algebraic loop" may emerge as follows. Consider $H : \mathcal{Z} \times \mathcal{V} \to \mathcal{W}$ as the output function of an I/O automaton so that $w = H(z, v)$ holds. During an evaluation of the control loop of Fig. 4.1, $L_p(z'_p, w_p, z_p, v_p) = 1$ must hold for $\mathcal{N}_p$ whereas

$L_c(z_c', w_c, z_c, v_c) = 1$ must hold for $\mathcal{N}_c$. These equations imply that $w_p = H_p(z_p, v_p)$ and $w_c = H_c(z_c, v_c)$. Based on (4.3),

$$
\begin{aligned}
w_c &= H_c(z_c, w_p) \\
&= H_c(z_c, H_p(z_p, v_p)) \\
&= H_c(z_c, H_p(z_p, w_c))
\end{aligned}
\tag{4.4}
$$

shows that w_c depends on itself through H_c and H_p for a given state couple (z_c, z_p). Similarly, the algebraic loop of w_p is given by

$$
w_p = H_p(z_p, H_c(z_c, w_p)).
\tag{4.5}
$$

$$
(4.4) \text{ and } (4.5) \Rightarrow \begin{cases} L_p(z_p', w_p, z_p, w_c) &= 1 \\ L_c(z_c', w_c, z_c, w_p) &= 1 \end{cases}
\tag{4.6}
$$

reflects the situation where the plant and the controller should switch from the states z_p and z_c to the states z_p' and z_c' respectively. The tuple (w_p, w_c) triggering this transition also solves the system of equations. The following sets of solution $\tilde{\mathcal{W}}_c(z_c, z_p)$ and $\tilde{\mathcal{W}}_p(z_c, z_p)$ are fixed points of the algebraic loops (4.4) and (4.5). They solve (4.6) for a state couple (z_c, z_p):

$$
\begin{aligned}
\tilde{\mathcal{W}}_c(z_c, z_p) &= \{w_c \in \mathcal{W}_c : L_p(z_p', w_p, z_p, w_c) = L_c(z_c', w_c, z_c, w_p) = 1, \\
&\quad (z_c', z_p') \in \mathcal{Z}_c \times \mathcal{Z}_p, w_p \in \mathcal{W}_{ap}(z_c, z_p)\}
\end{aligned}
\tag{4.7}
$$

$$
\begin{aligned}
\tilde{\mathcal{W}}_p(z_c, z_p) &= \{w_p \in \mathcal{W}_p : L_p(z_p', w_p, z_p, w_c) = L_c(z_c', w_c, z_c, w_p) = 1, \\
&\quad (z_c', z_p') \in \mathcal{Z}_c \times \mathcal{Z}_p, w_c \in \mathcal{W}_{ac}(z_c, z_p)\}.
\end{aligned}
\tag{4.8}
$$

A control loop $\mathcal{N}_l$ is said to be *well-posed* if for every state combination $z_l = (z_c, z_p)^T$ there is a *unique* input combination $(v_c, v_p) = (\tilde{w}_p, \tilde{w}_c)$ resulting from (4.7) and (4.8). [10] proposed the concept of *weak well-posedness* which is adapted to the control design discussed here. In the case of the weak well-posedness it is sufficient to exclude the trivial solution $\emptyset$ from the solution of (4.7) and (4.8). A control loop is said to be *ill-posed* if any of the fixed points sets is empty. These concepts are summarized in the following Definition.

Definition 4.2 (Well-posedness). *A control loop which consists of a plant $\mathcal{N}_p$ and a feedback controller $\mathcal{N}_c$ is*

- *well-posed iff*

$$|\tilde{\mathcal{W}}_c(z_c, z_p)| = |\tilde{\mathcal{W}}_p(z_c, z_p)| = 1, \tag{4.9}$$

- *weakly well-posed iff*

$$|\tilde{\mathcal{W}}_c(z_c, z_p)| > 0 \wedge |\tilde{\mathcal{W}}_p(z_c, z_p)| > 0, \tag{4.10}$$

- *and ill-posed iff*

$$|\tilde{\mathcal{W}}_c(z_c, z_p)| = 0 \vee |\tilde{\mathcal{W}}_p(z_c, z_p)| = 0, \tag{4.11}$$

with $\tilde{\mathcal{W}}_c(z_c, z_p)$ and $\tilde{\mathcal{W}}_p(z_c, z_p)$ defined in (4.7) and (4.8).

Now the following statements link the blocking property of a control loop with the well-posedness introduced above:

(S1) **Nonblocking control loop**: A control loop which consists of a plant $\mathcal{N}_p$ and a feedback controller $\mathcal{N}_c$ is nonblocking if it is either well-posed w.r.t. (4.9) or weakly well-posed w.r.t. (4.10).

(S2) **Blocking control loop**: A control loop which consists of a plant $\mathcal{N}_p$ and a feedback controller $\mathcal{N}_c$ is blocking whenever it is ill-posed w.r.t. (4.11).

Statement (S1) holds because in both situations of well-posedness and weak well-posedness, the controller and the plant always generate output events for their input events.
In statement (S2), ill-posedness emerges because either the plant, the controller or both cannot perform a conjoint state transition from their respective state z_c and z_p. According to Definition 4.1 there is no output sequence from the plant $\mathcal{N}_p$ which can be evaluated by the controller $\mathcal{N}_c$ in a way that it responds with an input sequence to the plant so that both components perform state transitions, i.e, from the considered state, they both satisfy (4.2) for any input sequence $V_p(0 \cdots k_e) \in \mathcal{V}_p^{k_e+1}$.
The characteristic function of the nonblocking control loop model L_l can never vanish during the considered horizon $0 \cdots k_e$. In that case, the following holds:

$$\bigvee_{z_p'}^{\mathcal{Z}_{ap}(z_p)} \bigvee_{w_l}^{\mathcal{W}_{ap}(z_p)} L_l((z_c', z_p')^T, w_l, (z_c, z_p)^T) > 0 \tag{4.12}$$

$$\bigvee_{z_p'}^{\mathcal{Z}_{ap}(z_p)} \bigvee_{w_l}^{\mathcal{W}_{ap}(z_p)} \bigvee_{v_p}^{\mathcal{V}_p} L_c(z_p', v_p, z_p, w_l) \cdot L_p(z_p', w_l, z_p, v_p) > 0 \ . \tag{4.13}$$

Note that it is assumed for simplicity that the states of the controller and those of the plant share the same labels z_p and z_p'. Otherwise, it would be necessary to add $\bigvee$ operators for the states of $\mathcal{N}_c$.

Remark   Condition (4.10) is less conservative than in [2] which requires that $|\breve{\mathcal{W}}_c(z_c, z_p)| = 1$ holds. Now (4.10) also includes W-deterministic maximally permissive controller $\mathcal{N}_c$ for which there exists a state z_c where (2.47) and $|\mathcal{V}_{ac}(z_c)| > 1$ hold. Fig. 4.3 shows the special case where the controller generates a unique control output sequence $(1, 2)$, despite the different measurements $(2, 4)$ or $(3, 5)$ from the plant to reach state 4.

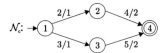

Figure 4.3.: Example of a W-deterministic controller with $|\mathcal{V}_{ac}(z_c)| > 1$

4.2. Specification of a process

4.2.1. Different types of specifications

A specification $\mathcal{S}$ describes how the plant $\mathcal{N}_p$ *should* behave under the influence of the controller $\mathcal{N}_c$. $\mathcal{S}$ is represented as follows by means of the "models" symbol $\models$:

- $\mathcal{S} \models Z_s$: The plant should follow a given state sequence $Z_s(0 \cdots k_e) = z_{0p} \times Z_s(1 \ldots k_e)$ where z_{0p} is given by $\mathcal{N}_p$.

- $\mathcal{S} \models z_F$: The plant should reach the final state z_F from the initial state z_{0p}. This specification is equivalent to the *marked state* concept widely used in automata theory.

- $\mathcal{S} \models W_s$: The plant should generate a given output sequence $W_s(0 \cdots k_e)$.

- $\mathcal{S} \models \mathcal{Z}_{ill}$: The plant should avoid a given set $\mathcal{Z}_{ill}$ of illegal (forbidden) states.

- $\mathcal{S} \models \mathcal{T}_{ill}$: The plant should avoid a given set of illegal I/O transitions $\mathcal{T}_{ill}$.

- $\mathcal{S} \models \mathcal{W}_{ill}$: The plant should not generate a given set of forbidden outputs $\mathcal{W}_{ill}$.

The specification types $\mathcal{Z}_{ill}$, $\mathcal{T}_{ill}$, and $\mathcal{W}_{ill}$ are suitable for, e.g., safety specifications in order to exclude dangerous states, transitions or outputs, respectively. Note that a specified final state z_F, a state sequence $Z_s(0 \cdots k_e)$ or an output sequence $W_s(0 \cdots k_e)$ can always express a safety specification type of forbidden states $\mathcal{Z}_{ill}$, forbidden transitions $\mathcal{T}_{ill}$ and forbidden outputs $\mathcal{W}_{ill}$. Therefore, only the first three specification types listed above, i.e, $\mathcal{S} \models Z_s$, $\mathcal{S} \models z_F$ and $\mathcal{S} \models W_s$ are considered in the following. Moreover, the method developed here does not consider mixing specifications of different types for the control design.

4.2.2. Interpretation of the specification automaton

In previous technical reports of the author [13], [14], the specification automaton $\mathcal{N}_s$ is defined as a subautomaton of the plant automaton $\mathcal{N}_p$. It is computed by means of the specification operator $\mathrm{Spec}(\cdot)$ which deletes every transition of $\mathcal{N}_p$ that is forbidden by the specification $\mathcal{S}$, so that

$$\mathcal{N}_s = \mathrm{Spec}(\mathcal{N}_p, \mathcal{S}) \subseteq \mathcal{N}_p \tag{4.14}$$

holds. However, the specification automaton obtained so far *contains* the behavior specified by $\mathcal{S}$ but does not reflect it exactly. This problem occurs whenever there is a state z_p in $\mathcal{N}_p$ with allowed transitions by the specification $\mathcal{S}$, so that

$$|\mathcal{Z}_{ap}(z_p)| > 1 \tag{4.15}$$

holds.

Example   Consider the plant described in Fig. 4.4 and the specification $\mathcal{S} \models Z_{s1}^*(0 \dots 3) = (1, 2, 4, 2)^*$. The resulting specification automaton of Fig. 4.5 is able to achieve the given state sequence but can also perform other state sequences like $Z_{s1\sharp}^*(0 \dots 3) = (1, 2, 1, 2)^*$. Hence, the resulting specification automaton of Fig. 4.5 obtained by deletion of forbidden transitions contains $\mathcal{S}$ but does not exactly reflect it.

The example shows that there is no guarantee that the specification can be executed by the plant although it is possible, i.e, feasible in $\mathcal{N}_p$ w.r.t [1]. For this reason such a specification automaton will be called *pseudo-specification automaton* (PSA) in the following.

Definition 4.3 (Pseudo-specification automaton of the plant $\mathcal{N}_p$). *A pseudo-specification automaton (PSA) $\mathcal{N}_s^{PSA} \subseteq \mathcal{N}_p$ is the largest subautomaton resulting from a deletion of every transition (z_p', w_p, z_p, v_p) in a plant $\mathcal{N}_p$ which does not satisfy the specification $\mathcal{S}$.*

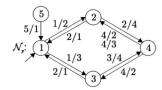

Figure 4.4.: Example of a plant automaton

It has to be studied if it is possible to derive a controller $\mathcal{A}_c$ based on a PSA to enforce the required behavior of $\mathcal{S}$ in $\mathcal{N}_p$. The existence of such a controller is referred to as the controllability of a specification in [2].

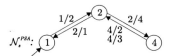

Figure 4.5.: Example of a pseudo-specification automaton (PSA)

Definition 4.4 (Exact specification automaton of the plant $\mathcal{N}_p$). *An exact specification automaton (ESA) $\mathcal{N}_s^{ESA}$ is an I/O automaton which reflects the sequential execution of a specification $\mathcal{S}$ in a plant $\mathcal{N}_p$.*

Fig. 4.6 shows a specification automaton which exactly reflects the given specification. Contrary to Fig. 4.5, every necessary transitions to achieve $Z_{s1}^*(0 \ldots 3)$ can be performed in $\mathcal{N}_s$ without ambiguity. Thus, such a specification automaton will be given the name *exact specification automaton* (ESA). The main difference concerns the fact that the state 2 has to be splitted into the states 2_1 and 2_2.

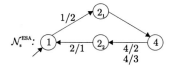

Figure 4.6.: Example of an exact specification automaton (ESA)

From a set theoretic viewpoint, it can be said that if a map between a set of specifications of the same type $\mathcal{S} = \{\mathcal{S}_1, \ldots, \mathcal{S}_\alpha\}$ and a set of specification automata $\mathcal{N}_s =$

$\{\mathcal{N}_{s1}, \ldots, \mathcal{N}_{s\beta}\}$ is considered, the map $\mathrm{Spec}_{\mathrm{PSA}}(\cdot)$ is allowed to be surjective (see Fig. 4.7). That is, several specifications may lead to the same specification automaton through $\mathrm{Spec}_{\mathrm{PSA}}(\cdot)$. Whereas in the case of ESAs, the $\mathrm{Spec}_{\mathrm{ESA}}(\cdot)$ map has to be injective (see Fig. 4.8).

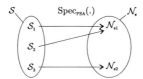

Figure 4.7.: Surjective specification operator for PSA design

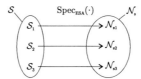

Figure 4.8.: Injective specification operator for ESA design

4.2.3. Pseudo-specification automaton

This structure refers to the one previously called specification automaton in [1, 4, 13] and [14]. The advantages and disadvantages are discussed next.

- Advantages of a PSA compared to an ESA

    1. The complexity of the PSA is bounded by the one of the plant $\mathcal{N}_p$ because of (4.14).

    2. The resulting controller can be realized in an Instruction List w.r.t. to the IEC-1131 standard [101].

    3. No renaming or splitting procedure of the states is needed, thus, they keep their meaning even after the DECD.

    4. The feasibility analysis of a specification can be carried out by means of the test automaton introduced in Section 4.2.5. For instance, an empty test automaton would describe an unfeasible specification.

- Disadvantages of a PSA compared to an ESA

  1. The difficulty of the computation of a homomorphism from $\mathcal{N}_s^{PSA}$ to the plant model $\mathcal{N}_p$ for the feasibility test.

  2. The determinism of $\mathcal{N}_s^{PSA}$ is indirectly visible, namely in interaction with the controller. It requires a specific realization architecture of the controller [2].

The first disadvantage emerges from the fact that it is often difficult to find a homomorphism for two well distinguished input/output automata. But in this case, as proven in [1] it is possible to consider the identity map $\mathcal{I}$ as a homomorphism which always exist because of (4.14). Moreover, an alternative feasibility analysis can be achieved with the test automaton used in Section 4.2.5 for the design of the PSA as mentioned in the 4th advantage of PSAs.

4.2.4. Exact specification automaton

This structure introduced in Section 4.2.2 must exactly reflect the specification $\mathcal{S}$. Its advantages and disadvantages are discussed in this section.

- Advantages of an ESA compared to a PSA

  1. The control design method is simplified to an inversion of the inputs and outputs of $\mathcal{N}_s^{ESA}$. The resulting controller has a well-suited form for sequential control or an Sequential Function Chart (SFC) w.r.t. to the IEC-1131 standard [101].

  2. The feasibility test of a specification is easier than for a PSA.

  3. The determinism of $\mathcal{N}_s^{ESA}$ is in general directly visible. Hence, no particular realization architecture of the controller is needed since the resulting controller always has the structure of a sequential functional chart where the states would be the steps and the I/O transition the switching condition.

- Disadvantages of an ESA compared to an PSA

  1. Arbitrary growing complexity with real and imaginary states resulting from the states splitting. The problem becomes worse for specifications with several cycles, e.g., for $Z_{s2}(0 \cdots k_e) = (1, 2, 4, 2, 4, 2, 4, 3, 1, 3, 1)$ in Fig. 4.4, 7 imaginary states have to be added to the 4 existing ones due to the splitting procedure. 5 additional transitions have to be defined for $\mathcal{N}_{s2}^{ESA}$. See the intermediate exact specification automaton of Fig. 4.9(a).

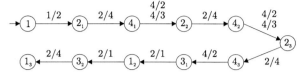

(a) Example of an intermediate exact specification automaton with steps as subscript

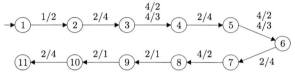

(b) Example of an exact specification automaton with renamed states

Figure 4.9.: Exemplary design of a an exact specification automaton

2. A renaming technique to distinguish real and imaginary states is needed. States information gets lost because they then become meaningless as in Fig. 4.9(b). Indeed, the new states just represent computing steps. Hence, the condition (4.14) does no longer hold.

3. ESAs are unpracticable not only for state sequences Z_s with degrees of freedom, i.e., a set of allowed states $\mathcal{Z}_{sk}$ at a given step k, but also for certain z_F or W_s specification types.

Note   This attempt of having an exact specification automaton is inspired by the language based automata theory where the specification automaton is already given by a specification language. Only the labels of the transitions, i.e, the events are relevant whereas the states are meaningless. The states are represented by an incremented counter which describes a particular string or a sequence of events. This is the same as the ESA introduced above. Hence, the renaming procedure makes the modeling efforts useless because a state which was assigned a specific meaning can now be expressed by several strings because of cycles.

4.2.5. Specification automaton design

The main idea is to derive an autonomous deterministic automaton $\mathcal{M}_s$ called *test automaton* from the specification. The test automaton consists of a characteristic function $\lambda_s(\cdot)$ which describes a set of state sequences $\mathcal{Z}_s(0 \cdots \boldsymbol{K_e})$. The characteristic function λ_s is built by ignoring the I/Os but solely considering the specified state transition sequences $\mathcal{Z}_s(0 \cdots \mathbf{K_e})$. For a state couple (z_s', z_s), $\lambda_s(z_s', z_s) = 1$ iff the transition (z_s', z_s) appears as a pair of consecutive states in at least one state sequence of $\mathcal{Z}_s(0 \cdots \mathbf{K_e})$. Note that $\mathcal{M}_s$ can be nondeterministic.

The product of the characteristic functions of $\mathcal{N}_p$ and $\mathcal{M}_s$ returns the value of the characteristic function of $\mathcal{N}_s$. These steps are summarized as follows:

1. Extract the state sequences set $\mathcal{Z}_s(0 \cdots \mathbf{K_e})$ out of $\mathcal{S}$

$$\mathcal{S} \implies \mathcal{Z}_s(0 \cdots \mathbf{K_e}) \tag{4.16}$$

   as described in Section 4.2.6.

2. Derive the characteristic function λ_s of the test automaton $\mathcal{M}_s$ from $\mathcal{Z}_s(0 \cdots \mathbf{K_e})$ as described in Section 4.2.7.

3. Compute the characteristic function of $\mathcal{N}_s$ with

$$\boxed{L_s(z_p', w, z_p, v) = L_p(z_p', w, z_p, v) \wedge \lambda_s(z_p', z_p).} \tag{4.17}$$

First, the state sequence set extraction procedure are formally presented now for the specification types Z_s, z_F and W_s.

4.2.6. State sequences set extraction

This section presents the extraction procedure of the state sequences set $\mathcal{Z}_s(0 \cdots \mathbf{K_e})$ for the state sequence specification $\mathcal{S} \models Z_s$, the final state specification $\mathcal{S} \models z_F$ and the output sequence specification $\mathcal{S} \models W_s$.

$\mathcal{Z}_s(0 \cdots \mathbf{K_e})$ **extraction for** $\mathcal{S} \models Z_s$. The naive solution in this case would be to say $\mathcal{Z}_s(0 \cdots \mathbf{K_e}) = \{Z_s(0 \cdots k_e)\}$. However, the former would not guarantee that any transition which is not possible in $\mathcal{N}_p$ but erroneously given by the system designer would be tracked down and rejected, so that $\mathcal{Z}_s(0 \cdots \mathbf{K_e}) = \emptyset$. This is ensured by the following equation which accepts $Z_s(0 \cdots k_e)$ only if every transition is feasible in $\mathcal{N}_p$:

$$\mathcal{Z}_s(0 \cdots \mathbf{K_e}) = \bigtimes_{k=0}^{k_e} \{Z_s(k) \cap \mathcal{Z}_{ap}(Z_s(k-1))\} \text{ with } \mathcal{Z}_{ap}(Z_s(-1)) = z_{0p}. \tag{4.18}$$

Note that, through the Cartesian product, (4.18) implies $\mathcal{Z}_s(0 \cdots \boldsymbol{K_e}) = \{Z_s(0 \cdots k_e)\}$ if the state sequence is feasible or $\mathcal{Z}_s(0 \cdots \boldsymbol{K_e}) = \emptyset$ otherwise.

$\mathcal{Z}_s(0 \cdots \boldsymbol{K_e})$ **extraction for** $\mathcal{S} \models z_F$. The main differences to other specification types concern the state sequences that are consistent with z_F. They are based on the following facts:

- Every state sequence $Z_{si}(0 \cdots k_e) \in \mathcal{Z}_s(0 \cdots \boldsymbol{K_e})$ ends in z_F, i.e, $Z_{si}(k_e) = \{z_F\}$, $i = 1 \ldots |\mathcal{Z}_s(0 \cdots \boldsymbol{K_e})|$.

- The state sequences $Z_{si}(0 \cdots k_e) \in \mathcal{Z}_s(0 \cdots \boldsymbol{K_e})$ must not have the same length as it is the case when $\mathcal{S} \models Z_s$ or $\mathcal{S} \models W_s$.

In order to simplify the nomenclature, it is assumed that z_F is reachable from z_{0p}. Let

$$\overline{\boldsymbol{A}}_z = \bigcup_{k=1}^{|\mathcal{Z}_p|-1} \boldsymbol{A}_z^k \tag{4.19}$$

be the symbolic reachability matrix of $\mathcal{N}_p$. Recall the following:

1. The $\cup$ operation in (4.19) is similar to the common matrix addition.

2. The matrix entry $\overline{\boldsymbol{A}}_z(i,j)$ contains every possible state sequence from state i to state j within at most $|\mathcal{Z}_p| - 1$ steps.

Hence, the state sequences set is obtained by

$$\mathcal{Z}_s(0 \cdots \boldsymbol{K_e}) = \overline{\boldsymbol{A}}_z(z_{0p}, z_F). \tag{4.20}$$

For instance, based on the state set $\mathcal{Z}_p = \{0, 1, 2, 3, 4, 5\}$, the entry $\overline{\boldsymbol{A}}_z(1, 4)$ contains every possible trace from the $1st$ state, i.e, $z_p = 0$ to the $4th$ state, i.e, $z_p = 3$ in $\mathcal{N}_p$ (Fig. 3.3), page 51, e.g., $(0, 1, 2, 1, 2, 3)$.

$\mathcal{Z}_s(0 \cdots \boldsymbol{K_e})$ **extraction for** $\mathcal{S} \models W_s$. Contrary to the previous specification types, the difficulty of this one emerges from the fact that it is no longer sufficient to consider only one transition to determine if it should belong to $\mathcal{M}_s$ or not. Instead, a succession of transitions that are consistent with $W_s(0 \cdots k_e)$ have to be considered. The set of all possible state sequences $\mathcal{Z}_s(0 \cdots k_e + 1)$ which can be visited while fulfilling $\mathcal{S} \models W_s(0 \cdots k_e)$ are computed as follows:

1. Build the set $\mathcal{Z}_j$ of the indices j of the adjacency matrix entries $A_w^{k_e+1}(z_{0p}, j)$ where $W_s(0 \cdots k_e)$ is an element:

$$\mathcal{Z}_j = \{j \in \{1, \ldots, |\mathcal{Z}_p|\} : W_s(0 \cdots k_e) \in A_w^{k_e+1}(z_{0p}, j)\}. \qquad (4.21)$$

2. Compute the set $\mathcal{Z}_s(0 \cdots K_e)$ with

$$\mathcal{Z}_s(0 \cdots K_e) = \{Z_s(0 \cdots k_e + 1) \in A_z^{k_e+1}(z_{0p}, j), \forall j \in \mathcal{Z}_j : \\ W_s(0 \cdots k_e) \in \underset{k=0}{\overset{k_e}{\times}} \mathcal{W}_{ap}(Z_s(k+1), Z_s(k))\}. \qquad (4.22)$$

4.2.7. Test automaton computation from a set of state sequences

Definition 4.5 (Test automaton). *For a given set of state sequences $\mathcal{Z}_s(0 \cdots K_e)$, the test automaton $\mathcal{M}_s$ is an autonomous automaton according to Definition 2.1 with $\mathcal{M}_s = (\mathcal{Z}_s, \lambda_s, z_{0s})$ where*

$$\exists z_s \in \mathcal{Z}_s : \sum_{z_s'}^{z_s} \lambda_s(z_s', z_s) > 0. \qquad (4.23)$$

The computation of the characteristic function $\lambda_s(\cdot)$ is explained in this section. For a state sequences set $\mathcal{Z}_s(0 \cdots K_e)$ the characteristic function of $\mathcal{M}_s$ is determined by

$$\lambda_s(z_s', z_s) = \overset{\mathcal{Z}_s(0 \ldots K_e)}{\underset{Z_{si}}{\bigvee}} \overset{|Z_{si}|-2}{\underset{k=0}{\bigvee}} [(Z_{si}(k) \equiv z_s) \wedge (Z_{si}(k+1) \equiv z_s')] \forall (z_s', z_s) \in \mathcal{Z}_p^2. \qquad (4.24)$$

If $\mathcal{Z}_s(0 \cdots K_e) = \emptyset$ w.r.t. (4.18), then $\lambda_s(z_s', z_s) = 0 \; \forall (z_s', z_s) \in \mathcal{Z}_p^2$ when (4.24) is applied. In that case the test automaton does not exist and the specification is, therefore, unfeasible in $\mathcal{N}_p$. The states set is given by

$$\mathcal{Z}_s = \{z \in \mathcal{Z}_p, \exists z' \in \mathcal{Z}_p : \lambda_s(z', z) \vee \lambda_s(z, z') = 1\} \qquad (4.25)$$

and $z_{0s} = \mathcal{Z}_s(0)$ under the assumption that $|\mathcal{Z}_s(0)| = 1$. Equations (4.24) and (4.25) are not only applicable for the specification type $\mathcal{S} \models Z_s$ but also for the types $\mathcal{S} \models z_F$ and $\mathcal{S} \models W_s$. In the sequel, the state sequences set extraction is explained on the following example.

Example. Consider the I/O automaton graph depicted in Fig. 3.3 which describes the fluid level control process presented in Section 1.3.1.

The specification is given by $Z_s = (0, (1, \ldots, 4, \ldots, 2)^*)$. The following state sequence satisfies this specification:

$$\mathcal{S} \models Z_s(0 \cdots 7) = (0, 1, 2, 3, 4, 3, 2, 1). \tag{4.26}$$

By applying (4.18), $\mathcal{Z}_s(0 \cdots \boldsymbol{K_e}) = \mathcal{Z}_s(0 \ldots 7) = \{Z_s(0 \ldots 7)\} \neq \emptyset$. The latter will be used in Section 4.2.8 to study the feasibility of the specification.

The characteristic function of the test automaton is obtained by (4.24) as follows

$$\lambda_s(z'_s, z_s) = \bigvee_{k=0}^{8-2} [(Z_s(k) \equiv z_s) \wedge (Z_s(k+1) \equiv z'_s)], \ \forall (z'_s, z_s) \in \mathcal{Z}_p^2. \tag{4.27}$$

For instance, $\lambda_s(2, 3) = [Z_s(5) \equiv 3] \wedge [Z_s(6) \equiv 2] = 1$. Once every transition has been processed with (4.24), the characteristic function λ_s of the test automaton $\mathcal{M}_s$ is ready to be multiplied with L_p to obtain the characteristic function L_s of the specification automaton $\mathcal{N}_s$ according to (4.17). In this case, every transition which is inconsistent with (4.26) is deleted from $\mathcal{N}_p$. The result is the specification automaton $\mathcal{N}_s$ depicted in Fig. 4.10. It contains the expected behavior of the plant under control. Thus, it is sufficient to use this specification automaton to design the controller. It is not necessary to have a specification automaton which exactly reflects $\mathcal{S}$. Since the aim of this chapter is to design a controller, it is sufficient to have every transition of $\mathcal{S}$ included in $\mathcal{N}_s$.

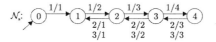

Figure 4.10.: Specification automaton graph of a fluid level control process

4.2.8. Feasibility of a specification

There are three kinds of criteria which have been developed to test the feasibility of a specification:

1. Homomorphism from $\mathcal{N}_s$ to $\mathcal{N}_p$ through the identity map $\mathcal{I}$ [1].

2. Formalism based on characteristic functions, set theory and trellis [5].

3. Adjacency matrices.

Compared to the homomorphic identity map and the adjacency matrices, the second test method based on characteristic functions, set theory and trellis is preferred. It has the advantage to be more rigorous and easily implementable. Therefore, the test of feasibility of a specification used in this thesis relies on the aforementioned formalism.

Definition 4.6 (Basic feasibility). *For a plant $\mathcal{N}_p$, a specification $\mathcal{S} \models Z_s$, $\mathcal{S} \models z_F$ or $\mathcal{S} \models W_s$ is basically feasible iff the test automaton $\mathcal{M}_s$ exists.*

Thus, the basic feasibility is fulfilled for specifications which *can be achieved* by the plant $\mathcal{N}_p$. This is equivalent with the fact that the set of state sequences $\mathcal{Z}_s(0 \cdots K_e)$ is nonempty as expressed by the following lemma.

Lemma 4.1. *For a given plant $\mathcal{N}_p$, a specification $\mathcal{S} \models Z_s$, $\mathcal{S} \models z_F$ or $\mathcal{S} \models W_s$ is basically feasible iff $\mathcal{Z}_s(0 \cdots K_e) \neq \emptyset$ w.r.t (4.18), (4.20) or (4.22) respectively.*

Proof. See Appendix B, page 207. □

Since the plant is modeled by a nondeterministic automaton, the property of basic feasibility ensures that the plant *can* achieve the specification with a controller. It does not guarantee that the plant *will always achieve* the specification with that controller. The requirement that the plant must always achieve the specification is called *safety*, similarly to [30, 113].

Definition 4.7 (Safe feasibility). *For a plant $\mathcal{N}_p$, a specification $\mathcal{S} \models Z_s$, $\mathcal{S} \models z_F$ or $\mathcal{S} \models W_s$ is said to be safely feasible if it is basically feasible and no input sequence $V_s(0 \ldots k)$ with $k > 0$ can also lead to another state sequence $Z_{s\sharp} \neq Z_s$, another final state $z_{F\sharp} \neq z_F$ or a another output sequence $W_{s\sharp} \neq W_s$, respectively.*

The basic feasibility condition from [1] is based on the existence of a homomorphism from $\mathcal{N}_s$ to $\mathcal{N}_p$. This property is extended here to the *safe feasibility* which means that there exists a controller $\mathcal{N}_c$ that will never block with $\mathcal{N}_p$ and enforce $\mathcal{S}$ to be achieved in the closed-loop. The safe feasibility is now presented for $\mathcal{S} \models Z_s$, $\mathcal{S} \models z_F$ and $\mathcal{S} \models W_s$.

Theorem 4.1 (Safe feasibility for $\mathcal{S} \models Z_s$). *A specification $Z_s(0 \cdots k_e)$ is safely feasible in the plant $\mathcal{N}_p$ iff*

1. Z_s is basically feasible for $\mathcal{N}_p$ w.r.t. Lemma 4.1 and

2. *the plant can not deviate from Z_s due to its nondeterminism, which means*

$$\bigvee_{k=0}^{k_e-1} \bigvee_{z_p'} \bigvee_{w_p} \bigvee_{v_s}^{Z_p \setminus z_{sk}' \; W_p \; V_{op}(z_{sk}';z_{sk})} L_p(z_p', w_p, z_{sk}, v_s) = 0 \tag{4.28}$$

with $z_{sk} = Z_s(k)$ and $z_{sk}' = Z_s(k+1)$.

Proof. See Appendix B, page 208. □

Equation (4.28) of Theorem 4.1 requires that the characteristic function of the plant vanishes for all transitions which may violate the specification $S \models Z_s$. These are sequences with critical states where the controller would send a command v_s for which the plant can move to both specified state z_{sk}' and a wrong state $z_p' \in \mathcal{Z}_p \setminus \{z_{sk}'\}$.

This concept is easily extended to an output sequence specification type $S \models W_s$ as in the Corollary 4.1. The extension consist of considering those state transitions $\mathcal{Z}_s(0 \cdots K_e)$ obtained through (4.22) which are consistent with $S \models W_s$. The characteristic function must vanish for output sequences which might also be consistent with state transitions of $\mathcal{Z}_s(0 \cdots K_e)$ but differ from W_s because of a wrong output $w_p \in \mathcal{W}_p \setminus w_{sk}$ at a given step k.

Corollary 4.1 (Safe feasibility for $S \models W_s$). *A specification $W_s(0 \cdots k_e)$ is safely feasible in the plant $\mathcal{N}_p$ iff*

1. *W_s is basically feasible for $\mathcal{N}_p$ w.r.t. Lemma 4.1 and*

2. *the plant can not generate different output sequences than W_s due to its nondeterminism, which means*

$$\bigvee_{k=0}^{k_e-1} \bigvee_{z_p'} \bigvee_{w_p} \bigvee_{v_s}^{Z_p \setminus z_{sk}' \; W_p \setminus w_{sk} \; V_{op}(z_{sk}';z_{sk})} L_p(z_p', w_p, z_{sk}, v_s) = 0,$$

$$\forall Z_s(0 \cdots k_e + 1) \in \mathcal{Z}_s(0 \cdots K_e) \text{ from (4.22)} \tag{4.29}$$

with $z_{sk} = Z_s(k)$, $z_{sk}' = Z_s(k+1)$, and $w_{sk} = W_s(k)$.

The remaining specification type for which the safe feasibility condition needs to be investigated is the final state specification $S \models z_F$.

Theorem 4.2 (Safe feasibility for $\mathcal{S} \models z_F$). *A specification $\mathcal{S} \models z_F$ is safely feasible for the plant $\mathcal{N}_p$ iff every state sequence $Z_s(0 \cdots k_e) \in \mathcal{Z}_s(0 \cdots K_e)$ is safely feasible by applying (4.28).*

Proof. See Appendix B, page 208. □

Note that, Theorem 4.2 is a generalization of Theorem 4.1 because it considers a set of state sequences contrary to the latter which is focussed on single state sequences. It requires that every state trajectory which can be used by the controller to drive the plant into the final state z_F be free of critical states from which a control input would lead to a state out of the considered trajectory. However, it is also conservative because the control loop may block in the case where the plant switches from a given state sequence $Z_{s1} \in \mathcal{Z}_s(0 \cdots K_e)$ to another state of sequence $Z_{s2} \in \mathcal{Z}_s(0 \cdots K_e)$ with $Z_{s1} \neq Z_{s2}$. However, such a behavior will be identified here as a fault because it would not be in accordance with the predicted behavior of the control loop by the diagnoser. Indeed, a controller executing a control law which stems from Z_{s1} would block if the plant begins to follow the trajectory Z_{s2} unless the I/Os of both specified state sequences are similar until the final state z_F.

4.2.9. Completeness and exactness of the specification

A *complete* and *exact* specification is presented in [119] as an ideal specification which should be given by the system designer. These two properties of a specification are now put in the context of this thesis.

Completeness. A complete specification demands to specify the behavior of the plant for all possible input sequences. Since it is practically difficult to obtain such a specification, a negatively formulated requirement will be used here. It states that any input which is not in line with the specification should be ignored (rejected) by the plant and the controller. The rejection of a bad input is reflected by a stagnation of the controller and the plant in their current state, respectively. Consequently, a faulty sensor will be ignored by the controller and must not be included in advance in the specification by the designer just for the need of completeness of the specification. This will be adopted in the realization of the controller in Section 4.3.3. Note that a complete specification including the faulty events which can flow through the control loop is close to the field of robust fault-tolerant control which is not the focus of this thesis.

Exactness. From a nominal control design point of view, it is appealing to define an exact specification to guarantee the determinism of the controller and reduce its complexity in certain cases. From a fault-tolerant control view point, it is obvious that the more exact a specification is defined, the less tolerant the nominal controller would be and the more sensitive the diagnosis unit would be. Therefore, degrees of freedom in the specification are necessary to demonstrate fault tolerance. The topic of degrees of freedom and redundancies is developed in Section 5.2.

4.3. Feedback control design

4.3.1. Main idea

The specification $\mathcal{S}$ is represented by the specification automaton $\mathcal{N}_s$. The main idea of the control design procedure is to keep the same structure of the specification automaton $\mathcal{N}_s$ but to reverse the input/output events v_s/w_s to get the controller automaton $\mathcal{N}_c$, i.e, $v_c/w_c = w_s/v_s$. To keep the same structure of the specification automaton means to use the same states label and state transitions. Formally, the main idea is summarized as follows:

$$\forall \, (z'_s, w_s, z_s, v_s) \in \mathcal{Z}_s \times \mathcal{W}_s \times \mathcal{Z}_s \times \mathcal{V}_s, z_c = z_s, z'_c = z'_s, v_c = w_s, w_c = v_s. \qquad (4.30)$$

However, this straightforward approach does not reveal how to derive a control law and how to explicitly enforce a control output to the plant. Moreover, a method describing how to get a controller out of a set of control laws is necessary and addressed in the following.

4.3.2. Procedural description of the design method

Possible control design procedures

A control law $\mathcal{A}_c^{(i)}$ ($i = 1 \ldots \nu$) is introduced as a subgraph of the supergraph $\mathcal{N}_c$, in the sense of [92], with a transition degree of one for all state couples through which the specification is achieved in the closed-loop with the plant $\mathcal{N}_p$. Since $\mathcal{A}_c^{(i)} \subseteq \mathcal{N}_c$ holds $\forall i \in [1, \nu]$ and $\mathcal{A}_c^{(i)} \neq \mathcal{A}_c^{(j)}$ for $i \neq j$, $\mathcal{N}_c$ is obtained by the union of the control laws $\mathcal{A}_c^{(i)}$:

$$\mathcal{N}_c = \bigcup_{i=1}^{\nu} \mathcal{A}_c^{(i)}. \qquad (4.31)$$

$\mathcal{N}_c$ is, therefore, called supercontroller in the following. An important property of $\mathcal{N}_c$ is that it is maximally permissive by construction. A supercontroller $\mathcal{N}_c$ may consists of several control laws $\mathcal{A}_c^{(i)}$. This chapter proposes two controller design flows:

a) **The maximally permissive controller synthesis:** the control laws $\mathcal{A}_c^{(i)}$ are derived from $\mathcal{N}_s$ and the specification $\mathcal{S}$. Then the maximally permissive controller $\mathcal{N}_c$ is obtained by merging all the control laws $\mathcal{A}_c^{(i)}$ together (see Fig. 4.11(a)).

b) **Supercontroller decomposition into control laws:** First, the maximally permissive controller $\mathcal{N}_c$ is obtained by inverting the inputs and outputs of $\mathcal{N}_s$. Then $\mathcal{N}_c$ is decomposed into several control laws $\mathcal{A}_c^{(i)}$ according to the specification $\mathcal{S}$ (see Fig. 4.11(b)).

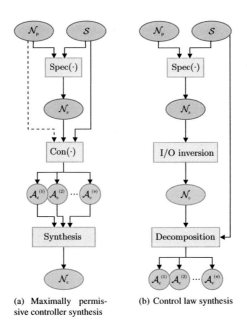

(a) Maximally permissive controller synthesis

(b) Control law synthesis

Figure 4.11.: Control design procedures

Since the specifications $\mathcal{S} \models z_F$ and $\mathcal{S} \models W_s$ can be expressed by sets of state sequences Z_s, the latter is used in the following as a canonical specification type for the control design.

Maximally permissive controller synthesis

For a specified state sequence $Z_s(0 \cdots k_e)$ which must be enforced by the controller, the number of possible control laws ν is given by

$$\nu = \prod_{k=0}^{k_e-1} \sum_{w_i}^{\mathcal{W}_{as}(z'_{sk}, z_{sk})} \sum_{v_i}^{\mathcal{V}_{as}(z'_{sk}, z_{sk})} L_s(z'_{sk}, w_i, z_{sk}, v_i) \tag{4.32}$$

$$= \prod_{k=0}^{k_e-1} td_s(z'_{sk}, z_{sk}) \tag{4.33}$$

where $z'_{sk} = Z_s(k+1)$ and $z_{sk} = Z_s(k)$. Equations (4.32) and (4.33) computes the product of the number of possible transitions for each state combination (z'_{sk}, z_{sk}) along the sequence $Z_s(0 \cdots k_e)$.

Example. For the specification automaton shown in Fig. 4.10, (4.32) yields

$$\begin{aligned}
\nu &= \prod_{k=0}^{6} td_s(z'_{sk}, z_{sk}) \\
&= td_s(1,0) \cdot td_s(2,1) \cdot td_s(3,2) \cdot td_s(4,3) \cdot \\
&\quad td_s(3,4) \cdot td_s(2,3) \cdot td_s(1,2) \\
&= 1 \cdot 1 \cdot 1 \cdot 1 \cdot 2 \cdot 2 \cdot 2 = 8.
\end{aligned} \tag{4.34}$$

Thus, there are 8 control law automata $\mathcal{A}_c^{(i)}$ with $i = 1 \ldots 8$. This is due to the fact that it is possible to use either valve V_2 or valve V_3 in 3 different states namely $2, 3$ and 4.

The design of an I/O controller automaton $\mathcal{N}_c = (\mathcal{Z}_c, \mathcal{V}_c, \mathcal{W}_c, L_c, z_{0c})$ consists of the following steps:

1. Build the specification automaton $\mathcal{N}_s$ according to (4.17).

2. Find the control laws $\mathcal{A}_c^{(1)}, \ldots, \mathcal{A}_c^{(\nu)}$ by means of

$$\mathcal{A}_c^{(i)} = \mathrm{Con}(\mathcal{N}_s, \mathcal{S}) \text{ with } i = 1 \ldots \nu, \tag{4.35}$$

   where the $\mathrm{Con}(\cdot)$ operator represents Algorithm 1.

3. Build the feedback controller $\mathcal{N}_c$ with the characteristic function

$$\boxed{\begin{aligned}
&L_c(z'_c, w_c, z_c, v_c) = \bigvee_{i=1}^{\nu} L_c^{(i)}(z'_c, w_c^{(i)}, z_c, v_c^{(i)}) = 1 \\
&\text{with } w_c \in \{w_c^{(i)}, i = 1 \ldots \nu\} \text{ and } v_c \in \{v_c^{(i)}, i = 1 \ldots \nu\}.
\end{aligned}} \tag{4.36}$$

$\mathcal{N}_c$ may become nondeterministic even though it fulfills $\mathcal{S}$. This is acceptable only if the nondeterminism of $\mathcal{N}_c$ solely concerns the internal state transition of the plant but not the outputs generation. That is, $\mathcal{N}_c$ must be at least W-deterministic w.r.t Lemma 2.1. However, the W-determinism of $\mathcal{N}_c$ does not guarantee the uniqueness of the control output.

The control design problem (4.35) is now derived from Section 4.1.1 for the first and third requirements as follows:

- Given:

    - A plant model $\mathcal{N}_p$

    - A specification $\mathcal{S}$ expressed by the canonical state sequence $Z_c(0 \cdots k_e)$

- Find: A set of controllers $\mathcal{A}_c^{(i)}$ ($i = 1 \ldots \nu$) which generate a unique input to the plant at each step k while executing $Z_c(0 \cdots k_e)$.

Since the control law $\mathcal{A}_c^{(i)}$ represents a unique subgraph describing a state sequence and a unique output sequence through the controller $\mathcal{N}_c$, the $\mathrm{Con}(\cdot)$ operator consists of a central routine which guarantees the uniqueness of these subgraphs. This routine generates all possible control laws $\mathcal{A}_c^{(i)}$ related to a specified state sequence $Z_c(0 \cdots k_e)$. It is represented in Algorithm 1.

Algorithm 1 Control design for a state sequence specification.

Input: $\mathcal{N}_s$, $Z_c(0 \cdots k_e)$
Init: $i = 1$, compute ν with (4.32), $\mathcal{VW}_{z'z} = [\,]$
  1: **for** $k = 0$ to $k_e - 1$ **do**
  2:    $z = Z_c(k),\, z' = Z_c(k+1)$
  3:    $\mathcal{VW}_{as_{z'z}} = \mathcal{V}_{as}(z', z) \times \mathcal{W}_{as}(z', z)$
  4:    $\mathcal{VW}_{z'z} = \{(v, w) \in \mathcal{VW}_{as_{z'z}} : L_s(z', w, z, v) = 1\}$
  5:    **for each** $(\tilde{v}, \tilde{w}) \in \mathcal{VW}_{z'z}$ **do**
  6:      $i = 1$
  7:      **while** $i \leq \nu$ **do**
  8:        **if** $L_c^{(i)}(z', w, z, v) == 0,\ \forall (v, w) \in \mathcal{V} \times \mathcal{W}$, **then**
  9:          $L_c^{(i)}(z', \tilde{v}, z, \tilde{w}) = 1,\, i = i + |\mathcal{VW}_{z'z}|$
10:       **else**
11:          $i = i + 1$
12:       **end if**
13:      **end while**
14:    **end for**
15: **end for**
16: **Output:** $\mathcal{A}_c^{(i)}$ ($i = 1 \ldots \nu$)

The control design operator $\text{Con}(\cdot)$ described in Algorithm 1 generates control laws $\mathcal{A}_c^{(i)}$ for a given state sequences Z_c that are given for the considered specification type. The following explains how to specify such state sequences:

- $\text{Con}(\mathcal{N}_s, z_F)$: find all state sequences $Z_c \in \mathcal{Z}_s(0 \cdots \boldsymbol{K_e})$ in $\mathcal{N}_s$ from z_{0s} to z_F by means of (4.20).

- $\text{Con}(\mathcal{N}_s, Z_s)$: find the longest state sequence $Z_c \subseteq Z_s$ with at most one cycle.

- $\text{Con}(\mathcal{N}_s, W_s)$: find all state sequences $Z_c \in \mathcal{Z}_s(0 \cdots \boldsymbol{K_e})$ which are consistent with W_s by means of (4.22).

After Algorithm 1 has generated all possible control laws $\mathcal{A}_c^{(i)}$, the maximally permissive controller $\mathcal{N}_c$ is obtained by (4.36).

Supercontroller decomposition into control laws

The I/O inversion of $\mathcal{N}_s$ is formalized by

$$L_c(z_c', w_c, z_c, v_c) = L_s(z_c', v_c, z_c, w_c), \ \forall (z_c', w_c, z_c, v_c) \in \mathcal{Z}_s \times \mathcal{V}_s \times \mathcal{Z}_s \times \mathcal{W}_s. \quad (4.37)$$

The decomposition step consists of applying the following equations for every state transition of $\mathcal{N}_s$ to generate all control laws:

1. For $\mathcal{S} \models Z_s, \mathcal{S} \models z_F$ or $\mathcal{S} \models W_s$, derive $\mathcal{Z}_s(0 \cdots \boldsymbol{K_e})$, the set of possible state sequences to execute to achieve $\mathcal{S}$ w.r.t. (4.18), (4.20) and (4.22). However, recall that $\mathcal{S} \models Z_s$ is used as the canonical specification in the sequel.

2. Compute the number ν of control laws:

   a) Build the $(0, td)$-adjacency matrix $\boldsymbol{A}$ w.r.t. (2.39).

   b) Find the number of control laws with

   $$\nu = \sum_{Z_{si}(0 \cdots k_{ei})}^{\mathcal{Z}_s(0 \cdots \boldsymbol{K_e})} \prod_{k=0}^{k_{ei}-1} \boldsymbol{A}(Z_{si}(k+1), Z_{si}(k)). \quad (4.38)$$

3. Compute the control laws by decomposing $\mathcal{N}_c$ with Algorithm 2.

Example: computation of ν. Consider the example of Fig. 3.3 with the state set

$$\mathcal{Z}_p = \{0, 1, 2, 3, 4, 5\}.$$

According to the position of each state in $\mathcal{Z}_p$, the numerical adjacency matrix is

$$A = \begin{pmatrix} 3 & 2 & 0 & 0 & 0 & 0 \\ 1 & 1 & 2 & 0 & 0 & 0 \\ 0 & 1 & 1 & 2 & 0 & 0 \\ 0 & 0 & 1 & 1 & 2 & 0 \\ 0 & 0 & 0 & 1 & 1 & 2 \\ 0 & 0 & 0 & 0 & 1 & 2 \end{pmatrix} \tag{4.39}$$

and the state sequence (4.26) is translated into

$$Z_s(0 \cdots 7) = (1, 2, 3, 4, 5, 4, 3, 2).$$

Equation (4.38) returns

$$\begin{aligned} \nu &= \prod_{k=0}^{6} A(Z_s(k+1), Z_s(k)) \\ &= A(2,1) \cdot A(3,2) \cdot A(4,3) \cdot \ldots \cdot A(2,3) \\ &= 1 \cdot 1 \cdot 1 \cdot 1 \cdot 2^3 = 8 \end{aligned} \tag{4.40}$$

which is similar to (4.34).

The main steps of the Algorithm 2 can be summarized in the sentence "Distribute each input/output transition (z', w, z, v) of $\mathcal{Z}_s(0 \cdots K_e)$ among the control laws $\mathcal{A}_c^{(i)}$, $i = 1, \ldots, \nu$".

The procedure is now explained in detail:

1. Go through each transition (z, z') of Z_s and read the I/O transitions v/w to be distributed the control laws $\mathcal{A}_c^{(i)}$. (Lines 1-4)

2. Go through every completed control law obtained so far until the number of expected control laws for the considered transition is reached (Line 5)

3. If the transition (z', w, z, v) is not yet saved in the current control law $\mathcal{A}_c^{(i)}$, then save it with the corresponding index i, otherwise take the next control law by incrementing i. (Lines 7-10)

Algorithm 2 Decomposition of $\mathcal{N}_c$ into control laws $\mathcal{A}_c^{(i)}$

Input: $\mathcal{N}_c$, $Z_c(0 \cdots k_e)$, compute ν with (4.38)

  1: **for** $k = 0$ to $k_e - 1$ **do**
  2:    $z = Z_c(k)$, $z' = Z_c(k+1)$
  3:    $\mathcal{VW}_{z'z} = \{(v, w) \in \mathcal{VW}_{ac_{z'z}} : L_c(z', w, z, v) = 1\}$
  4:    $i = 1$
  5:   **while** $i \leq \nu$ **do**
  6:     **for** each $(v, w) \in \mathcal{VW}_{z'z}$ **do**
  7:      **if** $L_c^{(i)}(z', w, z, v) == 0$ **then**
  8:       $L_c^{(i)}(z', w, z, v) = 1$
  9:      **end if**
10:      $i = i + 1$
11:     **end for**
12:   **end while**
13: **end for**
14: **Output:** $\mathcal{A}_c^{(i)}, i = 1, \ldots, \nu$

An extended version of Algorithm 2 is presented in [9]. The formalism for Algorithm 2 is now presented. First two assumptions are made in order to make the formal representation digestible:

- For a given transition $(Z_s(k), Z_s(k+1))$ of a state sequence Z_s of $\hat{\mathcal{Z}}_s(0 \cdots \hat{k}_e)$, only the I/O combinations v/w for which $L_c(Z_s(k+1), w, Z_s(k), v) = 1$ are considered next.

- A transition $(Z_s(k+1), w, Z_s(k), v)$ is saved in $\mathcal{A}_c^{(i)}$ only if is not already saved i.e. $L_c^{(i)}(Z_s(k+1), w, Z_s(k), v) = 0$, so that double entries are excluded.

However, both assumptions are handled in Algorithm 2. For a given set of state sequences $\mathcal{Z}_s(0 \cdots K_e)$, a supercontroller $\mathcal{N}_c$ and the number of control laws ν, the control laws are obtained through the equation:

$$L_c^{(i)}(Z_s(k+1), w, Z_s(k), v) = \begin{cases} 1 & \forall \, Z_s(0 \cdots k_e) \in \mathcal{Z}_s(0 \cdots K_e), \forall k \in [0, |Z_s| - 1], \\ & (v, w) \in \mathcal{V}_c \times \mathcal{W}_c : \\ & (td_c^{(i)}(Z_s(k+1), Z_s(k)) = 1) \wedge 1 \leq i \leq \nu \\ 0 & else. \end{cases}$$

$$(4.41)$$

4.3.3. Realization of the feedback controller

The following realization is devoted to plants for which the specification is safely feasible in the control loop. For implementation purposes, the following requirements must be taken into account:

- The control laws $\mathcal{A}_c^{(i)}$ can be used only when the resulting control loop is strictly well-posed w.r.t. (4.9). This implies that the plant must be W-deterministic.

- The supercontroller $\mathcal{N}_c$ can be used when the resulting control loop is strictly or weakly well-posed w.r.t. (4.10). In this case the plant does not need to not be W-deterministic.

The latter is used here to explain the realization scheme because it offers a general solution. The problem to be solved when trying to realize a feedback controller is to find a structure which enforces a specific control output w_c foreseen by the controller $\mathcal{N}_c$ for the plant $\mathcal{N}_p$ so that it responds with an output w_p. The output $w_p = v_c$ received by the controller $\mathcal{N}_c$ triggers its state transition only if the control loop is at least weakly well-posed w.r.t. Definition 4.2. Otherwise the control loop is blocking. The controller realization scheme

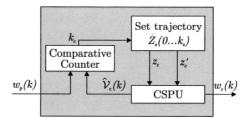

Figure 4.12.: Realization of the feedback controller

presented in Fig. 4.12 is summarized by the following equations:

- For a given final state z_F to be reached, the set trajectory is

$$\overline{Z}_s(0 \cdots k_e) \in \overline{A}_z(z_F, z_0). \tag{4.42}$$

- The counter k_c is a pointer along the set trajectory. It is incremented only if the measured output of the plant w_p matches with an expected value from $\hat{V}_c(k)$:

$$k_c = k_c + (w_p(k) \in \hat{V}_c(k)). \tag{4.43}$$

- The current state z_c and the current target state z_c' are consecutive elements of $\overline{Z}_s$, i.e.,

$$z_c = \overline{Z}_s(k_c) \text{ and } z_c' = \overline{Z}_s(k_c + 1). \tag{4.44}$$

- The control signal processing unit (CSPU) generates the control signal $w_c(k)$ and the expected outputs $\hat{\mathcal{V}}_c(k)$ from the plant to perform the state transition from z_c to z_c' and increment the counter k_c.

$$w_c(k) \in \mathcal{W}_{ac}(z_c', z_c) \text{ and} \tag{4.45}$$

$$\hat{\mathcal{V}}_c(k) = \mathcal{V}_{ac}(z_c', z_c). \tag{4.46}$$

4.3.4. Controllability conditions

W-**Determinism of the feedback controller.** This section proposes an answer to the question: When is the output generation of the controller deterministic? The following theorem from [1] states a criterion based on the specification automaton.

Theorem 4.3 (W-Determinism of $\mathcal{N}_c$)**.** *For a feasible specification $\mathcal{S}$, described by the automaton $\mathcal{N}_s$, there exists a W-deterministic controller $\mathcal{N}_c$ iff*

$$\forall (z_s, w_s) \in \mathcal{Z}_s \times \mathcal{W}_s, |\mathcal{V}_{as}(z_s, w_s)| = 1. \tag{4.47}$$

Proof. See Appendix B, page 209. □

The theorem holds obviously also for safely feasible specifications $\mathcal{S}$ since (4.47) concerns $\mathcal{N}_s$ only.

Controllability. The controllability of a plant for a given specification should describe the possibility to find a feedback controller with the requirements stated in Section 4.1.1 on page 61.

Definition 4.8 (Controllability of a plant)**.** *A plant $\mathcal{N}_p$ is said to be controllable w.r.t. a specification $\mathcal{S}$ iff there exists a W-deterministic feedback controller $\mathcal{N}_c$ for which $\mathcal{S}$ is fulfilled and the control loop is weakly well-posed.*

Since the first step of the control design method is to build the specification automaton $\mathcal{N}_s$, a necessary and sufficient condition for the existence of a controller $\mathcal{N}_c$ for a plant is the safe feasibility of the considered specification w.r.t Theorem 4.1, Corollary 4.1, and Theorem 4.2. However, the safe feasibility does not guarantee the W-determinism of the controller $\mathcal{N}_c$. Therefore, Theorem 4.3 gives a necessary and sufficient condition on the specification automaton under which it is possible to obtain a W-deterministic controller $\mathcal{N}_c$. Hence, it is an additional condition for the controllability of a plant. This is summarized by the following theorem.

Theorem 4.4 (Controllability). *A plant $\mathcal{N}_p$ is controllable w.r.t. a specification $\mathcal{S}$ iff $\mathcal{S}$ is safely feasible for $\mathcal{N}_p$ w.r.t Theorem 4.1 if $\mathcal{S} \models Z_s$, Corollary 4.1 if $\mathcal{S} \models W_s$ or Theorem 4.2 if $\mathcal{S} \models z_F$ and (4.47) holds.*

Proof. See Appendix B, page 209. □

The controllability condition proposed above is not based on the control laws $\mathcal{A}_c^{(i)}$ but on the controller $\mathcal{N}_c$ only. Note that a control loop $\mathcal{N}_c/\mathcal{N}_p$ will never block because (4.47) is fulfilled. On the contrary, a control loop $\mathcal{A}_c^{(i)}/\mathcal{N}_p$ may block because the plant may generate different outputs than the expected ones by $\mathcal{A}_c^{(i)}$, whereas this would not happen with $\mathcal{N}_c$ as a controller. This is the reason why control laws $\mathcal{A}_c^{(i)}$ are used only in strictly well-posed control loops for the controller realization (Section 4.3.3).

4.3.5. Complexity analysis

This section shows that the controller design problem is tractable when the main design procedures summarized in Algorithms 1 and 2 are analyzed. Firstly, the space complexity of the controller in both cases is obviously bounded, at worst, by the size of the state space $|\mathcal{Z}_p|$ of the plant. Secondly, a worst-case time complexity estimation is represented in the following by two functions $\mathcal{C}_1(n)$ and $\mathcal{C}_2(n)$ for the aforementioned algorithms, respectively. In both cases, the maximal number of input/output combinations $|\mathcal{V}_s| \cdot |\mathcal{W}_s|$ of the specification automaton $\mathcal{N}_s$ is represented by the variable $n \in \mathbb{N}$ in order to express the complexity of the problem. Thus, let $n = |\mathcal{V}_s| \cdot |\mathcal{W}_s|$ hold. In addition, assignments are assumed to require 1 execution time. Only relevant parts of the algorithms are explained next to justify the results:

1. Algorithm 1 (Control law design): $\mathcal{C}_1(n) \in \mathcal{O}(n^2)$. First the computation of ν w.r.t. (4.32) in the initialization step requires $k_e(|\mathcal{V}_s| + |\mathcal{W}_s|)$ executions in the worst case.

The first *for* loop in Line 1 is executed k_e times and the second one in Line 5 n times. The *while* loop in Line 7 is run $\nu + 1$ times. The *if* statement and the subsequent assignments require $n + 1$ steps (Lines 8-12). Based on these facts, the complexity of the whole algorithm is

$$\begin{aligned}
\mathcal{C}_1(n) &= 2 + k_e(|\mathcal{V}_s| + |\mathcal{W}_s|) + k_e[2 + 2n + n(1 + (\nu + 1)(n + 1))] \\
&= k_e(\nu + 1)n^2 + 4k_e n + (|\mathcal{V}_s| + |\mathcal{W}_s| + 2)k_e + 2. && (4.48) \\
\Rightarrow \mathcal{C}_1(n) &\in \mathcal{O}(n^2) && (4.49)
\end{aligned}$$

2. Algorithm 2 (Decomposition of $\mathcal{N}_c$): $\mathcal{C}_2(n) \in \mathcal{O}(n)$. The computation of ν w.r.t. (4.38) is not relevant in this algorithm because it is assumed to be given. However, for comparison purposes with Algorithm 1, note that it requires $\sum_{i=1}^{|K_e|} k_{ei}$ steps. Both *for* loops (Line 1 and Line 6) have the same complexity as in Algorithm 1. The *while* loop in Line 5 needs $2(\nu + 1)$ executions. The following complexity is deducted for this algorithm:

$$\begin{aligned}
\mathcal{C}_2(n) &= k_e[4 + n + 2(\nu + 1)(n + 4)] \\
&= k_e[(2\nu + 3)n + 8\nu + 12]. && (4.50) \\
\Rightarrow \mathcal{C}_2(n) &\in \mathcal{O}(n) && (4.51)
\end{aligned}$$

Equations (4.49) and (4.51) show that Algorithm 1 and 2 have a quadratic and a linear time complexity. Both estimations permit to conclude that the control design problem stated in Section 4.1.1 is deterministically tractable in polynomial time at worst.

Note that the complexity estimation given above can be improved in specific cases. In fact, $n = |\mathcal{V}_s| \cdot |\mathcal{W}_s|$ is an over-approximation which will rarely be necessary. It relies on the assumption that every input/output combination (v_s, w_s) of the specification automaton $\mathcal{N}_s$ is active at every state z_s. In the sense of [122], this is clearly a *pessimistic* estimation of the actual bound of the needed executions.

4.4. Application of the control design method

This section explains how the DECD method developed in this thesis can be applied for classic I/O automata and I/O trellis automata. Section 4.4.1 demonstrate the applicability of the control design approach in through simulations and experiments on a 3-Tank system. Section 4.4.2 highlights the benefits of I/O trellis for control design.

4.4.1. Nominal fluid level control process

Recall that the verbal specification of the process was formalized in (4.26) on page 77. The resulting specification automaton is depicted in Fig. 4.10, page 77.

According to (4.34) and (4.40), there are $\nu = 8$ control laws able to enforce (4.26) in $\mathcal{N}_p$. The first control law $\mathcal{A}_c^{(1)}$ obtained through Algorithm 2 is depicted in Fig. 4.13 as an example.

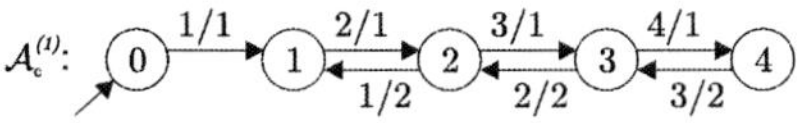

Figure 4.13.: Control law automaton $\mathcal{A}_c^{(1)}$ of the fluid level control process

The resulting maximally permissive controller $\mathcal{N}_c$ obtained by the subsequent application of Algorithm 1 and (4.36) is depicted in Fig. 4.14.

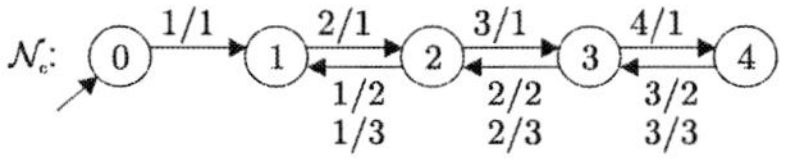

Figure 4.14.: Supercontroller of the fluid level control process

Simulation results. It is easy to verify that the behavior of $\mathcal{N}_p$ (Fig. 3.3) in a control loop with $\mathcal{A}_c^{(1)}$ (Fig. 4.13) fulfills the specification. Figure 4.15 illustrates the behavior of the nominal plant under control. It shows that the pump P_1 and the valve V_2 are alternatively activated the valve V_3 is unused in accordance with $\mathcal{A}_c^{(1)}$.

Experimental results. The control design method proposed here has been experimentally applied on the 3-Tank system depicted on Fig. 4.16 at the Institute of Automation and Computer Control (Ruhr-Universität Bochum). The experimental apparatus consists of 3 tanks with a maximal filling height of 0.5 m. They are equipped each with 5 differential pressure sensors and a manual outflow valve as e.g. V_3 (Fig. 4.16) to empty each tank down to the collecting tank TC. The tank T_2 is connected with T_1 and T_3 through electrically controlled valves V_2 which can be opened or closed. The pumps P_1 and P_2 are use to fill T_1 and T_3 with water from TC.

Actuator and sensor signals are connected with a PLC of Type SIMATIC C7-633 over a PROFIBUS communication. The PLC program exchanges those data with a personal computer (PC) in a 100 Mbit/s Ethernet network with a User Datagramm Protocol (UDP)

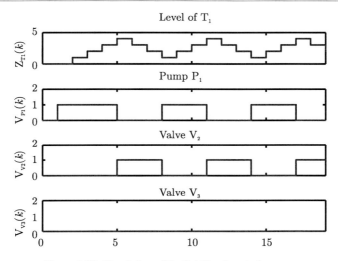

Figure 4.15.: Simulation of the fluid level control process

communication. The control law depicted in Fig. 4.13 is implemented w.r.t. Fig. 4.12 in MATLAB/Simulink with a sampling time of 0.5 s.

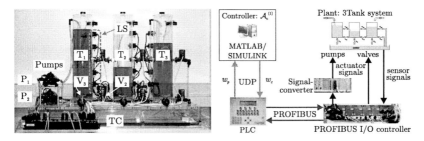

Figure 4.16.: Apparatus setup of the 3-Tank system for experimental fluid level control

In order to reflect the example used previously, the left tank T_1 is used. The manual valve of T_1 remains closed, the manual valve of T_2 remains open. The valve V_2 is opened or closed by the implemented controller depending on the sensor measurements LS and the next state to be reached according to the specification. The remaining actuators of the 3-Tank system are deactivated.

Figure 4.17 illustrates the behavior of the plant under control of $\mathcal{A}_c^{(1)}$ (Fig. 4.13). The first and second plots show how the controller alternates between the pump P_1 and the valve V_2 through $w_c = 1$ to fill T_1 and $w_c = 2$ to empty it. The third plot represents the set trajectory $\overline{Z}_s(0 \cdots k_e)$ of the controller as explained in Fig. 4.12. It is equivalent to the state sequence of the level of T_1 as specified by (4.26). The last plot shows the continuous evolution of the level of T_1.

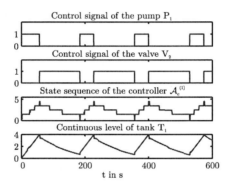

Figure 4.17.: Experimental realization of the fluid level control process

4.4.2. Control design method with I/O trellis automata

This section is intended to demonstrate the use of I/O trellis automata in the DECD method previously presented. Consider the model $\mathcal{N}_p$ of a plant exemplarily depicted by the non-deterministic I/O automaton in Fig. 4.18. It contains two possible initial states, cycles and redundant I/O transitions.

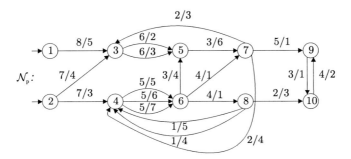

Figure 4.18.: Example of a plant model $\mathcal{N}_p$

The specification defined here for the plant $\mathcal{N}_p$ is $\mathcal{S} \models z_F = \{3\}$. The set of consistent state sequences $\mathcal{Z}_s(0 \cdots K_e)$ is obtained through (4.20) as follows:

$$
\mathcal{Z}_s(0 \cdots K_e) = \overline{A}_z(1,3) \cup \overline{A}_z(2,3) \tag{4.52}
$$
$$
= \{(1,3),(2,3),
$$
$$
(1,3,5,7,3),
$$
$$
(2,3,5,7,3),
$$
$$
(2,4,6,7,3),
$$
$$
(2,4,6,5,7,3),
$$
$$
(2,4,6,8,4,6,7,3),\ldots\}. \tag{4.53}
$$

$\mathcal{Z}_s(0 \cdots K_e)$ is used to derive a test automaton $\mathcal{A}_s$ w.r.t. (4.24). The resulting specification automaton of the plant $\mathcal{N}_p$ of Fig. 4.18 w.r.t the specification expressed by (4.53) is depicted in Fig. 4.19.

By applying the control design flow of Fig. 4.11(b) and (4.37), the supercontroller $\mathcal{N}_c$ is obtained as shown in Fig. 4.20. Its dynamic behavior is illustrated in the Trellis graph of Fig. 4.21 where the state 3 is in a bold node to highlight the beginning of new cycle according to the selected trajectories w.r.t to the specification expressed in (4.53). The I/Os are omitted for the sake of clarity. Note that the trellis graph also reflects the dynamic behavior of the test automaton $\mathcal{A}_s$ which stems from (4.53).

Some of the resulting control law automata $\mathcal{A}_c^{(1)}$ and $\mathcal{A}_c^{(2)}$ are depicted in Fig. 4.22 and fulfill the restriction $td(z',z) = 1 \ \forall (z',z) \in \mathcal{Z}_c^{(i)} \times \mathcal{Z}_c^{(i)}$ required in (4.41). The corresponding trellis graph in Fig. 4.23 illustrates the fact that $\mathcal{A}_c^{(1)}$ (dotted lines) and

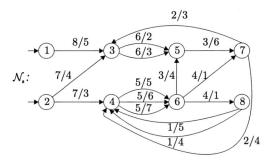

Figure 4.19.: Specification automaton of the plant model $\mathcal{N}_p$ from Fig. 4.18

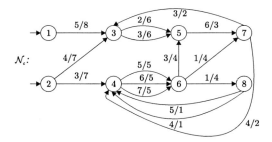

Figure 4.20.: Example of a supercontroller $\mathcal{N}_c$

$\mathcal{A}_c^{(2)}$ (dashed lines) are possible solutions of the control design problem considered in this example. Therefore, the application of Algorithm 2 on the supercontroller $\mathcal{N}_c$ works as expected.

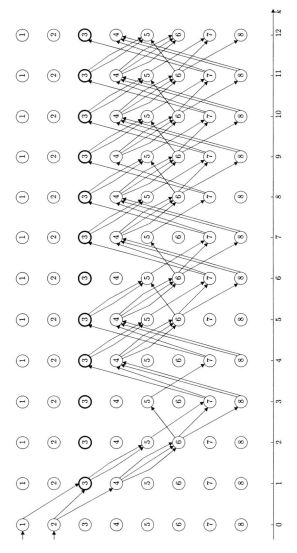

Figure 4.21.: Trellis graph of the supercontroller derived from Fig. 4.20

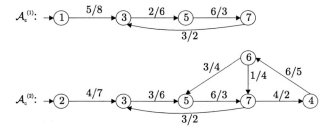

Figure 4.22.: Example of control laws $\mathcal{A}_c^{(i)}$ resulting from the decomposition of $\mathcal{N}_c$ of Fig. 4.20

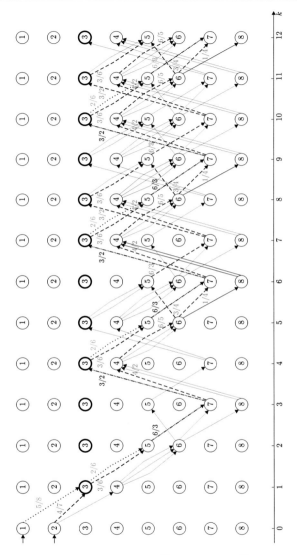

Figure 4.23.: Trellis graph of the control laws $\mathcal{A}_c^{(1)}$ (dotted) and $\mathcal{A}_c^{(2)}$ (dashed) derived from Fig. 4.22 as part of the supercontroller $\mathcal{N}_c$ in Fig. 4.21

4.5. Usability of the control design method

4.5.1. Implementation issues of the control law and control devices

Implementation issues. An encountered problem while trying to apply some DECD-techniques from the literature is the high-level of abstraction of the synthesized controllers or supervisors. Indeed, there is no direct (seldom simple) relationship between the synthesis solution and the real plant to be controled. Usually, additional information, manual steps and reformulations as well as a correct and adapted interpretation of the solution are necessary in order to obtain, e.g., a PLC-code.

From a practical point of view, the following requirements have to be considered to evaluate a formal control design method [125]:

- The model of the plant should be as detailed as possible but only as complex as needed. A compromise should be found here with fault-tolerant control requirements as mentioned in Section 2.3.2.

- The model of the plant has to be realized by an expert of the machine at hand.

- The model of the behavior of the system in the closed-loop must be easily representable.

- Analysis tools must exist to help engineers to verify or rethink their models and specification.

- Everything that is not leading to the control code must be automated.

The assumption that the I/O signals of a plant occur synchronously is suitable for theoretical issues since it simplifies the modeling and the simulation of a single system. If a closed-loop system is considered, this assumed synchronicity not only brings a fixed-point problem to deal with, but also renders the application on a real plant difficult. In reality, controllers and plants mostly behave asynchronously. In these cases the theoretical model of the control loop does not match the real control loop any more. For instance, the following situation can occur: The plant is at initial state z_0 and produces the output $w(0) = 0$ when switched on. In return, the controller produces the first control signal $v(0) = 1$, updates its internal states (z, z') with $(z(1), z(2))$ and expect to get the response $w(1) = 1$ that will activate the control signal $v(1)$, etc. From a theoretical point of view, the control loop is blocking for a certain time, namely as long as the plant does not generate the required output. Thus, it must be said that a blocking control loop is not necessarily a faulty one, the diagnosis has to decide. In practice, this short blocking period is not relevant in general. This is also what leads the discrete-event systems community to assume synchronicity as stated above.

Implementation devices. Possible control devices where the control method developed in this chapter can be applied are:

- Programmable Logic Controllers (PLCs): They are used this thesis as an interface between the controller implemented in MATLAB/Simulink and the plant (see Section 4.4.1 and Chapter 7).

- Field-Programmable Gate Arrays (FPGAs): They enable a rapid hardware reconfiguration as well as parallel signal processing. They have been used by, e.g., [70–72, 75] to implement learning algorithms as the backward propagation which requires recurrent reconfiguration of the FPGAs for given patterns. The use of FPGAs is also demonstrated in [164] to reconfigure neural networks.

- Microcontrollers: They often serve as embedded controllers where the discrete-event controller realization scheme can be implemented. Limitations concern the usually small memory amount available on microcontroller chips.

4.5.2. Application of the controller design approach to standard automata

First a conversion of $\mathcal{N}_p$ needs to be done by transforming every I/O transition v/w in an event sequence $v - w$. Inputs have to be controllable whereas outputs must be observable events as mentioned in Section 2.3.1. Then the specification also needs to be transformed as follows:

- Type z_F: declare z_F to be a marked state.

- Type Z_s: read the I/O sequences that are consistent with $Z_s(0 \cdots k_e)$, convert each of them in event sequences, derive all languages K_i from them, then perform a composition of those obtained languages.

- Type W_s: detect the input sequences related to $W_s(0 \cdots k_e)$, then derive all languages K_i from them, then perform a composition of those obtained languages.

- Type $\mathcal{Z}_{ill}, \mathcal{W}_{ill}, \mathcal{T}_{ill}$: delete the forbidden elements from $\mathcal{N}_p$, read all possible I/O sequences from the remaining transitions of $\mathcal{N}_p$, derive event sequences as above as specified languages K_i, then perform a composition to obtain the main specification.

Since the language K_i is usually already the supervisor, a direct way of computing the controller is to reverse the input-output sequences to obtain those the I/O automaton that would play the role of the controller. The resulting controller should be equivalent or at least isomorph to an ESA-based controller.

4.6. Literature review

Several approaches to control design exist in literature and a comparative overview is given in [152]. All these methods have the common goal that the controller is designed so as to suppress dangerous events or to suppress forbidden states in order to satisfy safety requirements. The differences to the approach proposed in this thesis lie in the way how the controller is derived for operating constraints and how the specification is modeled.

A widely used control design approach for standard automata was developed in [156], which is called in the following RW-Theory (RWT). Some RWT-based approaches for I/O automata were proposed by [42, 149] and [150]. Both references pointed to limitations of the RW-Theory for I/O automata, namely the lack of automatic synthesis, a poor implementation structure and the high computational complexity.

The approach proposed here contributes to the automatic synthesis of a controller for a given specification and provides sufficient information for implementation purposes. The computational complexity is a major problem of discrete-event systems in general and is not the concern of this thesis. However, the common issues and differences between the approach of this thesis and the RWT-based ones are highlighted in the following and motivate the development of a new approach.

The main difference between the I/O automata handled by RWT-based approaches and those presented here is the interpretation of I/O transitions. References [42, 56, 156] consider an I/O transition as a succession of an input event σ_i and an output event σ_o among 3 states (Fig. 4.24(a)). The I/O automata used in this framework are more compact because only one state transition and 2 states are taken into account for each I/O transition labeled v/w (Fig. 4.24(b)).

In addition, compared to standard automata, I/O automata explicitly describe the action-reaction principle (causality) which is a fundamental property of technological systems. The resulting I/O controller developed in [149] and [150] is a Moore automaton. In the

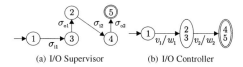

(a) I/O Supervisor   (b) I/O Controller

Figure 4.24.: Difference between a supervisor and a controller

framework of Fig. 4.1, this would lead to an open-loop control structure because the control signals w_c would be generated regardless of the outputs w_p of the plant, but would depend only on the internal state z_c of the controller due to the Moore property.

In the RW-Theory, the model of the control loop is represented by a synchronous product of the plant automaton with the supervisor. Verification techniques like model checking build a model of the plant and the violating specification invariant to check whether there is an event sequence that leads to a blocking situation ([39, 58] and [157]). This is similar to the violation of the feasibility condition proposed here by a dangerous state sequence. A critical discussion on the nonblockingness of control loops with RWT-based supervisors is given in [103].

The approach proposed in this thesis handles the plant $\mathcal{N}_p$ and the controller $\mathcal{N}_c$ as two well distinguished entities. Hence, the specification of the final state z_F in Fig. 4.1 is not designated for the control loop but for the plant only, but in a closed-loop with the controller. However, two key properties are required for the control loop: *determinism* and *nonblockingness*.

Another crucial difference is in the interpretation of the role of the controller. In the classic RWT, the controller is a supervisor which is supposed to enable or disable the controllable events of the plant in order to satisfy the specification. The enforcement of certain event is indirectly achieved by disabling others [82], by defining at most one controllable event at each state of the controller [170] or by introducing temporal conditions [121]. The objective of the controller developed here is to directly enforce a control input w_c to the plant in order to fulfill the specification. It is important to note that the feedback connection between the RWT-based supervisor and the plant is considered only for design purposes. In fact the obtained supervisor is not in an explicit feedback connection with the plant but is merged with the plant by using the synchronous product or the parallel composition. The automaton composed in this way is a generator with the language of the plant under control. This is the reason why a classical RWT-based supervisor is usually not ready to be used for I/O automata in its original version and cannot be directly used as a controller $\mathcal{N}_c$ in Fig. 4.1.

Other authors working with I/O automata are [43, 64, 194] and [148]. [194] address a general supervisory control problem where not only the interaction between the plant and the controller is considered but also their respective interaction with the environment.

[35] presents the control design problem for timed automata in terms of a game including the controller and the plant as the main players. The controller is said to have won the game, if it succeeds to drive the plant into a winning, i.e, accepting state.

Recently, [54] and [166] proposed an approach to extract a controller as a Mealy automaton out of a RWT-based supervisor. In order to use this approach in the context of I/O automata, it would be necessary to first convert I/O automata into standard automata, then to translate the specification into a language, build the supervisor and finally extract the needed controller. Instead, a straightforward method to get a controller which is ready to be used and easy to be implemented is proposed in this thesis. Resulting control laws can be implemented on a PLC as Statement List (STL), Ladder Logic (LAD), Function Block Diagram (FBD), Sequential Function Chart (SFC) and others. It is possible to switch among the description languages as explained in [157].

[105] study the problem of state avoidance whereas this thesis deals with the problem of reaching given states. It is possible to additionally put a requirement of avoiding a set of states $\mathcal{Z}_{ill}$. Those states will also be excluded while building the specification automaton.

Petri nets. [191] developed a widely approach of supervisor synthesis for Petri nets based on places invariants as explained in [138]. [96] give a survey of Petri nets based control design methods. Later on, [63, 104] presented a Petri net based approach with explicit inputs and outputs to enforce specific behaviors in the plant instead of just supervising (enabling and disabling) them as in the classical RW-Theory. This is similar to the enforcing concept proposed here. They make use of active signals which enforce specific behaviors and of passive signals to enable or prohibit some events. This approach is close to the one presented here for I/O automata. Since the controller structure of [104] is adapted to the number of I/Os of the plant, its size grows with the growing number of I/Os. In the approach presented here, the internal architecture of the controller is fixed and remains the same regardless of the size of the plant.

[125] classify DECD formalism into: Petri nets, Boolean algebra and Max-Plus-Algebra (see the references therein). Petri nets based structures called Net condition event systems (NCES) models are used to propose a formal control design method. Only reachable states are considered whereas the analysis tools mostly concern the reachability analysis and unfolding issues. The NCES are modules representing the behavior of a particular part of the plant. The specification is defined in terms of forbidden states and forbidden transitions.

The specification is then integrated in the NCES as a specific module. [125] also distinguish between passiv and active controllers. Supervisors are said to be passiv because they only enable or disable a fixed set of events of the plant whereas active controllers have the ability not only to enable or disable some events but also to enforce a specific behavior. The DECD as well as the analysis of closed-loops as investigated in [125] are applied in [106] through a Petri net markup language (PNML).

Comparison with model checking and verification. [52] and some references therein focus on the model checking problem. Since verification algorithms check all possible state sequences resulting from manual design, they usually involve more state sequences than controller synthesis algorithms as Algorithm 2 or Algorithm 1 (see [103]).

The basic feasibility condition of Definition 4.6 can be used to express the typical model checking problem. A dangerous state sequence specification Z_{s-ill} should not be feasible at all in $\mathcal{N}_p$. The corresponding set of state sequences $\mathcal{Z}_s(0 \cdots K_e)$ must then be empty as opposed to Lemma 4.1. That is, $\mathcal{N}_p$ should not be a homomorphic image of the resulting specification automaton $\mathcal{N}_{s-ill}$ via the identity map $\mathcal{I}$ as it is the case for required specification. This shows how this framework can be used to solve model checking problems once they are formulated correspondingly.

A crucial issue called the compatibility in [89] is the possibility of the existence of a controller able to drive the plant from an unknown initial set to a target sequence. This fits in the concept of feasibility of a specification developed here. Moreover, the causality imposed on the plant is a kind of implicit determinism, because the output sequences depend on the input history which is similar to a Mealy-property.

Model matching. The strong model matching problem for deterministic and completely defined I/O automata is studied in [43]. It consists of finding a controller for a given open-loop system with a desired closed-loop behavior. The controller synthesis proposed here is designed for the closed-loop (Fig. 4.1) even though it can run in an open-loop manner, e.g., if the controller fulfills the Moore property.

The references [79, 90, 91, 136, 137, 179, 180, 192, 193] and [148] develop their controller under the assumption of a deterministic control loop, whereas this thesis uses the notion of weak well-posedness to catch the nondeterminism of the plant with a deterministic controller. Furthermore, the existence condition of a controller is given in [79] by non empty entries in a boolean reachability matrix called skeleton matrix. In this thesis, the presence of a nonempty entry in such a reachability matrix is not sufficient to guarantee the achievement of the specification by the controller because of the nondeterministic behavior of the plant. The notion of safe feasibility was introduced to solve this issue. W-deterministic maximally permissive controllers introduced here can guarantee a safe achievement of the specification.

The difference between the control design problem stated here and the model matching problem is also in the reference model or the specification. The model matching problem uses a model of the control as the specification to fulfill, i.e, a controller that can lead to the specified control loop is to be found. In this thesis the controller should enforce a specification in the plant in a feedback connection and guarantee the safety and liveness of the control loop.

[133] proposed a systematic transformation technique, through algorithms, from a regular expression to a graph and vice-versa. The suggested framework is used there for specification modeling as it has been done later in the discrete-event systems community e.g [156].

5. Fault-tolerant control of discrete-event systems

Abstract. *The focus of this chapter is put on the reconfiguration unit of Fig. 1.1. This chapter explains how the dashed line in Fig. 1.1, which symbolizes the modification of the controller, has to be interpreted. The off-line and on-line reconfiguration concepts are presented first. The issues of degrees of freedom and redundancies which are crucial for fault tolerance are formalized. Necessary and sufficient conditions of reconfigurability are presented for actuator faults, sensor faults and system internal faults.*

5.1. Discrete-event control reconfiguration schemes

The difference between *off-line* and *on-line* reconfiguration resides in the means used to achieve a controller reconfiguration. The main differences are specified in the following subsections.

5.1.1. Off-line reconfiguration concept

The main characteristics of an off-line reconfiguration method are

- an interruption or even shut-down of the process for an indeterminate time,

- a global search of redundancies in the whole model of the faulty plant,

- a new control law able to permanently counteract the current failure.

The reconfiguration approach of [1] is suitable for the *off-line computation of a reconfigured control law* $\mathcal{A}_c^r$ to cope with actuator failures. In the example therein, the broken valve V_2 is replaced by the valve V_3 in the new control law to permanently counteract the actuator failure.

5.1.2. Formalized off-line reconfiguration

In case of a faulty actuator, the plant may either stop to work or continue to be in motion. The latter is possible due to a faulty command that is still active in the controller output memory or due to the natural dynamics of the plant in a given environment. Hence, in order to avoid possibly unsafe states, the current controller must be put off-line. This is symbolized by setting the plant input to ε. The formalization of two main reconfiguration techniques is presented now: the *trajectory re-planning* and the *Input/Output adaptation*.

Trajectory re-planning. The following steps need to be applied:

1. Build the model of the faulty plant $\mathcal{N}_p^f$ as in Section 3.6.4.

2. Copy the current control law and trajectory to be modified

$$\mathcal{A}_c^r = \mathcal{A}_c^{(i)} \tag{5.1}$$

$$\overline{Z}_c^r = \overline{Z}_c^{(i)}. \tag{5.2}$$

3. Delete the transitions from the control law where the input v_p^n, the output w_p^n or the next state z_p^n of the model of the nominal plant which lead to the blocking situation:

$$L_c^r(z', v_p^n, z, w) = 0, \forall (z', z, w) \in \mathcal{Z}_p^2 \times \mathcal{W}_p \tag{5.3}$$

$$L_c^r(z', v, z, w_p^n) = 0, \forall (z', v, z) \in \mathcal{Z}_p \times \mathcal{V}_p \times \mathcal{Z}_p \tag{5.4}$$

$$L_c^r(z_p^n, v, z, w) = 0, \forall (v, z, w) \in \mathcal{V}_p \times \mathcal{Z}_p \times \mathcal{W}_p. \tag{5.5}$$

4. Put the controller off-line and stop the own dynamic of the plant

$$\overline{Z}_c^r(k_i + 1) = z_\varepsilon \Rightarrow v_p(k_i) = w_c(k_i) = \mathcal{W}_{ac}(z_\varepsilon, *) = \varepsilon. \tag{5.6}$$

5. Synchronize the states of $\overline{Z}_c^r$ with $\mathcal{N}_p^f$

$$\overline{Z}_c^r(k_i) = z_p^f. \tag{5.7}$$

6. Find the smallest number of steps needed from z_p^f to z_p^n by solving the following equation for l:

$$l = \min\{l \in [1, |\mathcal{Z}_p^f|] : \mathbf{A}^l(z_p^n, z_p^f) > 0\}. \tag{5.8}$$

7. Iteratively find the set $\mathcal{Z}^r(k_i \cdots k_i + l)$ of state sequences from z_p^f to z_p^n within $l' = 1 \ldots l$ steps:

   $l' = 1$:

   $$\mathcal{Z}^r(k_i \cdots k_i + 1) = \{(z_p^f, z) \in (\mathcal{Z}_p^f)^2 : \boldsymbol{A}(z, z_p^f) > 0\}.$$

   $1 < l' \leq l$:

   $$\mathcal{Z}^r(k_i \cdots k_i + l') = \{(z_p^f, \ldots, z^-, z) \in \mathcal{Z}^r(0 \cdots l' - 1) \times \mathcal{Z}_p^f : \boldsymbol{A}^{l'}(z, z^-) > 0\}$$

   $$\Rightarrow \mathcal{Z}^r(k_i \ldots k_i + l) = \{(z_p^f, \ldots, z) \in \mathcal{Z}^r(0 \cdots l) : z = z_p^n\}. \tag{5.9}$$

8. Select one path Z^r from $\mathcal{Z}^r(k_i \ldots k_i + l) \setminus \mathcal{Z}_{ill}^r$ in $\overline{Z}_c^r$ from z_p^f to z_p^n, where $\mathcal{Z}_{ill}^r$ is the set of forbidden sequences and insert it into $\overline{Z}_c^r(k_i \cdots k_i + l)$.

9. Read the I/O signals in $\mathcal{N}_p^f$ along Z^r, inverse them and write them in $\mathcal{A}_c^r$ so that the following holds:

$$\boxed{\begin{array}{l} \bigwedge_{k=k_i}^{k_i+l-1} L_c^r(z', w, z, v) = 1 \text{ with } z' = Z^r(k+1), z = Z^r(k) \\ \text{and } L_p^f(z', v, z, w) = 1. \end{array}} \tag{5.10}$$

In case of an actuator failure, if for any $(z', z) \in (\mathcal{Z}_p^f)^2$ $v_p^n \in \mathcal{V}_{ap}^f(z', z)$ holds, then withdraw the sequence Z^r, add it to $\mathcal{Z}_{ill}^r$ and go back to step 8.

Input/Output adaptation. Adapt the nominal controller $\mathcal{A}_c$ according to the fault:

- Actuator faults: consider the actuator signal falsification due to $w_c^f = E_v(w_c^n)$. Thus, every controller output w_c^f must be replaced by the command w_c^r needed by the plant after the falsification $E_v(\cdot)$:

$$w_c^r \in \{w_c \in \mathcal{V}_p^f : w_c^n = E_v(w_c)\}. \tag{5.11}$$

The reconfigured characteristic function is given by

$$\boxed{L_c^r(z', w_c^r, z, v) = L_c^{(i)}(z', w_c^n, z, v) \; \forall z' \in \mathcal{Z}_c, z \in \mathcal{Z}_c, v \in \mathcal{V}_c.} \tag{5.12}$$

- Sensor faults: consider the sensor signal falsification due to $w_p^f = E_w(w_p^n) \neq \hat{v}_c^n$

where $\hat{v}_c^n$ is the input event expected by the controller. The reconfigured character-istic function is obtained by replacing every nominal input w_p^n of $\mathcal{A}_c^r$ by the falsified measurement $w_p^f = E_w(w_p^n)$ to hide the fault, i.e.,

$$\boxed{L_c^r(z', w_c, z, w_p^f) = L_c^{(i)}(z', w_c, z, \hat{v}_c) \ \forall z' \in \mathcal{Z}_c, z_c \in \mathcal{Z}_c, v \in \mathcal{V}_c} \qquad (5.13)$$

Limitations are encountered with this approach because the algorithm searches for us-able redundancies in the whole faulty plant model $\mathcal{N}_p^f$ in order to cope with the fault. As the computation of the reconfigured control law $\mathcal{A}_c^r$ requires the investigation of every tran-sition in the faulty plant model $\mathcal{N}_p^f$ regardless of their usability for the fault that occurred, this approach tends to be applicable only on systems with a large memory, computation resources and weak time constraints. For systems with hard real-time constraints where a reaction of the fault is required on-line, this approach is unapplicable. For the latter case, Section 5.1.3 proposes a suitable approach.

5.1.3. On-line reconfiguration concept

Opposed to the off-line-reconfiguration paradigm, the main characteristics of an on-line reconfiguration method are

- an interruption of the process for the shortest time possible in the ideal case,

- a local search of redundancies between the current state of the faulty plant and the next specified state only,

- a new control law able to temporarily, i.e., at a given step k, counteract the current fault.

The approach proposed in [4] is adequate for the *on-line computation of a controller output* $w_c(k_f + 1)$ adapted to a given fault. Should the effect of the fault go beyond the step $k_f + 1$, then another reconfiguration step will be executed for the same fault. In the example of the faulty fluid level control process, this would result in a stepwise replacement of valve V_2 by the valve V_3. This reconfiguration will take place until the resulting control law uses valve V_3 only. This stepwise computed control law would be the same as the one obtained through an off-line reconfiguration in a single run.

The on-line reconfiguration scheme proposed in this thesis is depicted in Fig. 5.1. The main parts are described separately in the following:

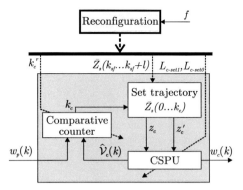

Figure 5.1.: On-line control reconfiguration scheme

1. The nominal controller mechanism represented by the lower colored block.

2. Explicit reconfiguration signals as dashed lines, namely:

   - the correction term k_c^r
   - the recovery subsequence $\bar{Z}_s(k_{sf} \ldots k_{sf} + l)$ and
   - the new characteristic function values of L_c to turn the nominal control law $\mathcal{A}_c$ into a reconfigured one $\mathcal{A}_c^r$.

3. The reconfiguration algorithm (Algorithm 3) represented by the upper colored block named correspondingly.

Since the on-line reconfiguration focuses only on relevant transitions, it is faster than an off-line reconfiguration. In this thesis, on-line reconfiguration is divided in the following strategies: *forward* and *backward* reconfiguration. Both concepts are first presented in a general sense in Section 5.1.4 and precisely explained in Section 5.3.

5.1.4. Forward and backward reconfiguration paradigm

The forward and backward reconfiguration strategies are said in [119] to depend on the ability of the reconfiguration unit to predict a damage or not. Predicted faults are said to be anticipated and unpredicted faults are said to be unanticipated. In that perspective, forward recovery is then dedicated to anticipated faults and backward recovery to unanticipated faults. Since any fault considered here is assumed to occur suddenly, it is always unantici-pated in the sense of [119]. Instead, its occurring step, as identified by the diagnosis unit, will be the main criterion to decide whether forward recovery or backward recovery should be applied.

An important contribution of this thesis is that a model-based implementation of the for-ward recovery is provided although it was claimed by [119] to be impossible. For backward recovery, no checkpoint are established in advance as in [31, 61, 67] and [119]. Instead, any reachable state prior to the fault is taken into account for the backward recovery devel-oped here. However, the following aspects stressed by [119] concur in a broad sense with this thesis:

- Backward recovery can always be applied regardless of the fault. Since an automated reconfiguration is pursued here, the latter statement is limited here by the physical constraints of the plant although it is always possible to reset the controller to any state.

- Forward recovery offers in general the better cost-effective options than backward recovery.

5.2. Degrees of freedom and redundancies

Although degrees of freedom and redundancies are both correlated, the former will be used here for specifications and the latter for controllers:

- $df(\mathcal{S})$ represents the degree of freedom of the specification $\mathcal{S} \models Z_s$, $\mathcal{S} \models z_F$ or $\mathcal{S} \models W_s$.

- $rd_W(z_c)$, $rd_T(z_c)$ represent the output-based and the transition-based redundancy degree, respectively, of a state z_c in a controller $\mathcal{N}_c$. $RD(\mathcal{N}_c)$ is the redundancy degree of a controller $\mathcal{N}_c$.

Both, the degrees of freedom and the redundancy degrees are formally presented in the subsequent sections.

5.2.1. Degrees of freedom of a specification

Degrees of freedom can be explicit or implicit. Explicit degrees of freedom consist of a set of states $\mathcal{Z}_k$ or outputs $\mathcal{W}_k$ at a particular step k within a state or output sequence. Implicit degrees of freedom are given, e.g., per definition in the specification of type $\mathcal{S} \models z_F$, since any state sequence from an initial state to the final state z_F fulfills the specification. Another implicit degree of freedom is a set of state (or output) sequences which are in accordance with an output (or state) sequence $Z_s(0 \cdots k_e)$ (or $W_s(0 \cdots k_e)$) apparently free of degrees of freedom. The latter does of course highly depend on the structure of the plant $\mathcal{N}_p$. The goal of the reconfiguration unit is to exploit those transitions with adequate (explicit or implicit) degrees of freedom to achieve the specification despite the faults.

Since it is possible to express each specification type, $\mathcal{S} \models Z_s$, $\mathcal{S} \models z_F$ or $\mathcal{S} \models W_s$, as a set of state sequences $\mathcal{Z}_s(0 \cdots \boldsymbol{K_e})$ by means of (4.18), (4.20), (4.22), respectively, $\mathcal{Z}_s(0 \cdots \boldsymbol{K_e})$ is used to formally define degrees of freedom for any of the three specification. Thus, the degree of freedom of a set of state sequences is defined as

$$df(\mathcal{Z}_s(0 \cdots \boldsymbol{K_e})) = \max_{k=0 \cdots \boldsymbol{K_e}} \{|\mathcal{Z}_{sk}|\} - 1. \tag{5.14}$$

Note that, if $|\mathcal{Z}_s(0 \cdots \boldsymbol{K_e})| = 1$, the specification $\mathcal{S} \models Z_s$, $\mathcal{S} \models z_F$ or $\mathcal{S} \models W_s$, has no degree of freedom because exactly one state is foreseen for the plant at every step k.

5.2.2. Redundancy degree of a controller

Since I/O automata are used to model the plant and the controller, two kinds of redundancies are presented:

- Transition-based redundancies which are intended to be used for models of the plant and

- Output-based redundancies which are intended to be use for the controller.

Input redundancies could similarly be defined but they are irrelevant from a controller view point. Note that output based redundancies are more specific than transition based ones which include redundant state, input and output transitions. These redundancies are typical for so called *multigraphs* [85], which are graphs in which for some states z and z' there is more than one transition. In the sequel, the redundancy degree of

- a state z_c and

- the supercontroller $\mathcal{N}_c$

are considered.

The specific case of control output-based redundancies for actuator failures is handled in [1] and now formalized as the number of alternative output sequences which are consistent with the set of state sequences $\mathcal{Z}_c(0 \cdots K_e)$ from an initial state to z_c.

Definition 5.1 (Output-based redundancy degree of a state). *For any given state z_c in $\mathcal{N}_c$, let $\mathcal{Z}_c(0 \cdots K_e)$ be the set of state sequences $Z_{ci}(0 \cdots k_{ei})$ with $Z_{ci}(0) = z_{0c}$ and $Z_{ci}(k_{ei}) = z_c$ with $i = 1 \ldots |\mathcal{Z}_c(0 \cdots K_e)|$. The output-based redundancy degree of a state z_c denoted $rd_W(z_c)$ is the number of alternative output sequences resulting from*

$$rd_W(z_c) = \sum_{Z_{ci}(0 \cdots k_{ei})}^{\mathcal{Z}_c(0 \cdots K_e)} \prod_{k=1}^{k_{ei}-1} \mathcal{W}_{ac}(Z_{ci}(k+1), Z_{ci}(k)) - 1 \qquad (5.15)$$

or

$$rd_W(z_c) = \left| \bigcup_{k=1}^{|Z_c|-1} A_w^k(z_{0c}, z_c) \right| - 1. \qquad (5.16)$$

Equation (5.16) requires the construction of the symbolic output adjacency matrix A_w in contrast to (5.15) which precisely selects the transitions to involve in the computation.

The definition for transition-based redundancies is presented next. It consists of finding the number of alternative paths from an initial state to a target state.

In order to compute the number of distinct paths eventually sharing transitions, solutions are suggested in [57, 140, 141] where the so called k-best path problem is addressed. The problem consists of finding a set of paths minimizing a given cost function based on the weights assigned to the transitions while only completely disjoint paths are considered. If the resulting set of paths is a singleton, then the problem is equivalent to the shortest path problem [120]. Reference [195] considers the minimum edges shareability, which it is not an issue to be discussed here. Since the cost of these paths are irrelevant in this thesis, the solution of [130] could be applied for transitions having the same costs, e.g., 1. However it is not adequate for the problem to be solved here, because the transitions are labeled with probabilities.

Recall that the elements of a $(0, td)$-adjacency matrix A reflect the number of transitions between each state couple (z', z). Hence, the entries $a_{z'z}$ of the matrix A^k reveals the number of state sequences $Z(0 \ldots k)$ from state $z = Z(0)$ to state $z' = Z(k)$ within k steps. It is important to note that the mentioned state sequences may contain internal cycles. Selfloops are not counted twice as often suggested in literature. This concept is exploited in the following definitions.

Definition 5.2 (Transition based redundancy degree of a state). *The transition-based redundancy degree of a state z_c is the maximal number of alternative distinct finite state sequences from any initial state z_{0c} to the given state z_c within at most $|\mathcal{Z}_c| - 1$ steps. Hence*

$$rd_T(z_c) = \max_{\substack{k=1\ldots|\mathcal{Z}_c|-1 \\ z_{0ci}\in\mathcal{Z}_c}} \{A^k(z_c, z_{0ci})\} - 1. \tag{5.17}$$

The redundancy degree of a state as in (5.17) permits to cover state and I/O redundancies from an initial to a target state. Compared to the output based redundancy degree, the transition-based redundancy degree of a state is easier to compute than its corresponding output based redundancy degree. Therefore, it can be used to determine the nonreconfigurability of a controller before attempting a reconfiguration.

Definition 5.3 (Redundancy degree of a supercontroller). *The redundancy degree of a supercontroller $\mathcal{N}_c$ is the maximal output based redundancy degree over every states of $\mathcal{Z}_c$. Hence*

$$RD(\mathcal{N}_c) = \max_{z_c\in\mathcal{Z}_c}\{rd_W(z_c)\}. \tag{5.18}$$

This redundancy degree of the supercontroller will be used to express the reconfigurability condition.

[173] proposes a definition of a so called strong redundancy degree which solely concerns the plant and not the controller as it is the case here. Another difference, with the approach presented here, is the fact that the redundancy degree in [173] is defined in relation with the fulfillment of a property by the system, e.g., the observability via some redundant sensors whereas the fulfillment of a sequence is relevant here. The similarity with this thesis lays in the interpretation of the redundancy degree. Namely, it reveals the maximal number of break downs or loss of transitions before the system loses the ability to ensure a certain property. To interpret the redundancy degree defined here in the sense of [173], the redundancy degree serves to measure the feasibility of a specification in a faulty control loop.

Example. To visualize the applicability of the redundancy definitions, consider the supercontroller exemplary depicted in Fig. 4.20, page 96. The numerical adjacency matrix is

$$
A = \begin{pmatrix}
0 & 0 & 0 & 0 & 0 & 0 & 0 & 0 \\
0 & 0 & 0 & 0 & 0 & 0 & 0 & 0 \\
1 & 1 & 0 & 0 & 0 & 0 & 1 & 0 \\
0 & 1 & 0 & 0 & 0 & 0 & 1 & 2 \\
0 & 0 & 2 & 0 & 0 & 1 & 0 & 0 \\
0 & 0 & 0 & 3 & 0 & 0 & 0 & 0 \\
0 & 0 & 0 & 0 & 1 & 1 & 0 & 0 \\
0 & 0 & 0 & 0 & 0 & 1 & 0 & 0
\end{pmatrix}.
$$

For instance, the entry $A(6,4) = 3$ reflects the transition degree of 3 from state 4 to state 6. Now, the computation of, e.g., A^5 to

$$
A^5 = \begin{pmatrix}
0 & 0 & 0 & 0 & 0 & 0 & 0 & 0 \\
0 & 0 & 0 & 0 & 0 & 0 & 0 & 0 \\
0 & 3 & 0 & 0 & 5 & 11 & 3 & 6 \\
0 & 3 & 0 & 0 & 11 & 29 & 3 & 6 \\
4 & 10 & 6 & 33 & 0 & 3 & 10 & 12 \\
6 & 33 & 0 & 9 & 0 & 0 & 33 & 54 \\
0 & 0 & 10 & 33 & 3 & 14 & 0 & 0 \\
0 & 0 & 6 & 27 & 0 & 3 & 0 & 0
\end{pmatrix}
$$

shows that there are 5 possible trajectories from state 5 to state 3 within 5 steps since $A^5(3,5) = 5$.

The transition-based redundancy degree $rd_T(3)$ is determined by computing the numerical adjacency matrix for the maximal horizon of steps for which any reachable state should be visited:

$$
A^7 = \begin{pmatrix}
0 & 0 & 0 & 0 & 0 & 0 & 0 & 0 \\
0 & 0 & 0 & 0 & 0 & 0 & 0 & 0 \\
10 & 43 & 6 & 42 & 0 & 3 & 43 & 66 \\
22 & 109 & 6 & 60 & 0 & 3 & 109 & 174 \\
0 & 9 & 20 & 66 & 39 & 115 & 9 & 18 \\
0 & 0 & 66 & 261 & 9 & 60 & 0 & 0 \\
6 & 48 & 0 & 9 & 43 & 109 & 48 & 84 \\
0 & 9 & 0 & 0 & 33 & 87 & 9 & 18
\end{pmatrix}
$$

Equation (5.17) leads to $rd_T(3) = 43$.

Since it is counter-intuitive to involve sequences already containing the target state in the enumeration of redundancies, some arguments for this decision are given. In fact, the concept of redundancy in its largest sense requires to consider every possible trace from an initial state to a target state including cycles as far as they are distinguishable. To exclude these cycles would be equivalent to removing redundant paths, thus, reducing redundancies. In addition, redundancies are needed for control reconfiguration as suggested in [1] and [4]. A redundant cycle would be needed, e.g., if $\mathcal{S} \models z_F = \{3\}$ and a fault takes the plant away from 3 to 5. The redundant cycle $(5, 7, 3)$ can be used by the reconfigured controller to achieve the specification.

Discussion. Since (5.18) is applicable on the plant $\mathcal{N}_p$, it is possible to derive a necessary condition for the reconfigurability of a plant. The condition would be formulated as follows. A control loop consisting of a plant $\mathcal{N}_p$ under the influence of actuator or internal state failure is reconfigurable if

$$RD(\mathcal{N}_p) > 0 \ w.r.t. \ (5.18).\qquad(5.19)$$

However, the methods developed here show that the intuitive statement of (5.19) is not always true, due to the following reasons:

- The fact that $RD(\mathcal{N}_p) > 0$ holds, does not guarantee the safe feasibility of a specification $\mathcal{S}$ proven in Theorem 4.1, Corollary 4.1 and Theorem 4.2.

- Even though $RD(\mathcal{N}_p) = 0$, the controller can still be reconfigured for actuator or sensor faults where the signals are falsified and the falsification relation Err_v or Err_w defined in Section 3.5.2 and Section 3.5.3, respectively, are known. If the fault is, e.g., a bit toggling of the inputs v_p or outputs w_p, then the controller can still be reconfigured despite the lack of redundancies in the plant w.r.t. (5.19).

5.3. Forward and backward on-line reconfiguration of a controller

Firstly, the main steps of Algorithm 3 are explained. It describes how the three key reconfiguration signals $k_c, \overline{Z}_s(k_{sf} \dots k_{sf} + l)$ and $L_c^r(\cdot)$ from Fig. 5.1 are computed on-line. It is assumed that a fault can be diagnosed within a finite number of steps delay. Algorithm 3 implements two recovering strategies:

- **Backward recovery** which is necessary when a fault is identified to have occurred in the past, that is $k_{sf} < k_s$. A transition is added from the current state of the plant $Z_p(k)$ back to a healthy state $\overline{Z}_s(k_{sf})$ of the controller, where the fault took place.

- **Forward recovery** which is applied when the fault is identified to have occurred when the controller is actually blocking due to an unexpected plant output $w_p(k_f)$. Since none of the expected input was received by the controller, they all have to be deleted from its control law regardless of the possible control outputs. That means to build L_{c-set0}.

Depending on the recovering strategy at hand, the corresponding state sequence $\overline{Z}_s(k_{sf} \ldots k_{sf} + l)$ is computed afterwards. This is done by means of a Breadth-First-Search algorithm labeled by the BFS$(\cdot)$ operator in a while loop. BFS$(z_{start}, z_{end}, \mathcal{N})$ searches for every state sequence denoted $\overline{Z}_{BFS}$ between z_{start} and z_{end} in the I/O automaton $\mathcal{N}$. When the result of the BFS$(\cdot)$ operator is empty ($\overline{Z}_{BFS} = \{\}$), its current z_{end} input parameter is added to $\overline{Z}_{s-skip}$ describing the set of states to be skipped during future rounds and the counter n is incremented. The same holds for computed state sequences $\overline{Z}_{BFS}$ for which the current output of the plant $w_p(k_f)$ would still not belong to the expected inputs $\mathcal{V}_{ac}(\overline{Z}_{BFS}(2), \overline{Z}_{BFS}(1))$ of the controller, i.e., these sequences are useless and must be deleted. The useful sequence $\overline{Z}_{BFS}$ needs to be inserted in the state sequence $\overline{Z}_s(0 \cdots k_e)$ of the controller without overwriting it completely. This leads to a new state sequence $\overline{Z}_s(k_s \ldots k_s + l)$ where

$$l = \begin{cases} n & \text{Forward recovery} \\ |\overline{Z}_{BFS}| + k_e - 2k_{sf} - n & \text{Backward recovery} \end{cases} \tag{5.20}$$

Note that the second equation of (5.20) results from

$$l = (|\overline{Z}_{BFS}| - 1 - k_{sf}) + (k_e - (k_{sf} + n) + 1).$$

The last step is to build L_{c-set1}, i.e, to set the values of the characteristic function of the transitions involved in $\overline{Z}_{BFS}$ to 1 and so turn the control law $\mathcal{A}_c$ into a reconfigured one: $\mathcal{A}_c^r$.

Algorithm 3 On-line control reconfiguration procedure

Input: $f, \mathcal{N}_p^f, k_{sf}, w_p(k_f), w_c(k_f), \mathcal{S}, \mathcal{A}_c, \overline{Z}_s(0 \cdots k_e)$

Init: $\overline{Z}_{\text{BFS}} = \{\}, k_c = 0, \overline{Z}_{s-skip} = \{\}, L_{c-set1} = L_{c-set0} = \{\}, n = 1, l = 0$

    // Strategy selection

1: **if** $k_{sf} < k_s$ **then** // Backward recovery
2:     $L_c(\overline{Z}_s(k_{sf}), \mathcal{V}_{ap}^f(\overline{Z}_s(k_{sf}), Z_p(k_f)), Z_p(k_f), \mathcal{W}_{ap}^f(\overline{Z}_s(k_{sf}), Z_p(k_f))) = 1$
3:     $k_c = -1$
4: **else** // Prepare forward recovery, build L_{c-set0}
5:     $L_c(\overline{Z}_s(k_{sf} + 1), *, \overline{Z}_s(k_{sf}), \mathcal{V}_{ac}(\overline{Z}_s(k_{sf} + 1), \overline{Z}_s(k_{sf}))) = 0$
6:     $L_{c-set0}(\overline{Z}_s(k_{sf} + 1), *, \overline{Z}_s(k_{sf}), \mathcal{V}_{ac}(\overline{Z}_s(k_{sf} + 1), \overline{Z}_s(k_{sf}))) = 0$
7: **end if**

    // Computation of $\overline{Z}_s(k_{sf} \ldots k_{sf} + l)$

8: **while** $w_p(k_f) \notin \mathcal{V}_{ac}(\overline{Z}_{\text{BFS}}(2), \overline{Z}_{\text{BFS}}(1))$ && $k_{sf} + k_c < k_e$ **do**
9:     **if** $\overline{Z}_s(k_{sf} + n) \in \overline{Z}_{s-skip} \,||\, (\overline{Z}_{\text{BFS}} \neq \{\} \,\&\&\, w_p(k_f) \notin \mathcal{V}_{ac}(\overline{Z}_{\text{BFS}}(2), \overline{Z}_{\text{BFS}}(1)))$ **then**
        // Delete useless sequences
10:        $\overline{Z}_{s-skip} = \overline{Z}_{s-skip} \cup \{\overline{Z}_s(k_{sf} + n), \overline{Z}_{\text{BFS}}(2)\}$
11:        $\overline{Z}_{\text{BFS}} = \{\}$
12:        $n = n + 1$
13:     **else**
14:        **if** $k_c == 0$ **then** // Forward recovery
15:           $\overline{Z}_{\text{BFS}} = \text{BFS}(\overline{Z}_s(k_{sf}), \overline{Z}_s(k_{sf} + n), \mathcal{N}_p^f)$
16:        **else if** $k_c == -1$ **then** // Backward recovery
17:           $\overline{Z}_{\text{BFS}} = \text{BFS}(Z_p(k_f), \overline{Z}_s(k_{sf} - n), \mathcal{N}_p^f)$
18:        **end if**
19:     **end if**
20: **end while**
21: $\overline{Z}_s(k_{sf} \ldots k_{sf} + l) = [\overline{Z}_{\text{BFS}}(k_{sf} \ldots |\overline{Z}_{\text{BFS}}| - 1), \overline{Z}_s(k_{sf} + n \ldots k_e)]$

    // Build L_{c-set1}

22: **for** $j = 1$ to $|\overline{Z}_{\text{BFS}}| - 1$ **do**
23:     $z = \overline{Z}_{\text{BFS}}(j), z' = \overline{Z}_{\text{BFS}}(j + 1)$
24:     $\mathcal{V}_a = \mathcal{V}_{ac}(z', z), \mathcal{W}_a = \mathcal{W}_{ac}(z', z)$
25:     **if** $L_c(z', \mathcal{W}_a(1), z, \mathcal{V}_a(1)) == 0$ **then**
26:        $L_{c-set1} = \left\{ \begin{array}{c} L_{c-set1} \\ (z', \mathcal{W}_a(1), z, \mathcal{V}_a(1)) \end{array} \right\}$
27:     **end if**
28: **end for**
29: **Output:** $k_c, \overline{Z}_s(k_{sf} \ldots k_{sf} + l), L_{c-set1}, L_{c-set0}$

5.4. Reconfigurability of a controller

This section presents the conditions under which the reconfiguration method, previously introduced, can be successfully achieved. The possibility to achieve a successful control reconfiguration is now defined as the reconfigurability.

Definition 5.4 (Reconfigurability). *A control loop is said to be reconfigurable if there exists a new control law $\mathcal{A}_c^r$ for the faulty plant $\mathcal{N}_p^f$ with the same specification $\mathcal{S}$ as in the nominal case.*

This property is now studied for actuator failures, actuator faults and sensor faults. The proofs of the following theorems remain in this chapter instead of being listed in Appendix B as the others. The reason is to highlight the overlapping between those theorems, especially the safe feasibility of the specification $\mathcal{S}$ in the faulty plant which represent a key condition in all the following theorems.

Theorem 5.1 (Reconfigurability for actuator failures). *For a specification $\mathcal{S}$, a control law $\mathcal{A}_c^{(i)} \subseteq \mathcal{N}_c$ and a faulty plant automaton $\mathcal{N}_p^f$ subject to an actuator failure, there exists a reconfigured controller $\mathcal{A}_c^r$ iff*

1. $RD(\mathcal{N}_c) > 0$ and

2. $\mathcal{S}$ is safely feasible in $\mathcal{N}_p^f$ according to Theorem 4.1, Corollary 4.1 or Theorem 4.2.

Proof of Theorem 5.1. ($\Longrightarrow$) If $\exists \mathcal{A}_c^r = (\mathcal{Z}_c^r, \mathcal{W}_c^r, \mathcal{V}_c^r, L_c^r, z_{0c}^r)$, then $\mathcal{Z}_c^r$ and the other sets are non empty. Let $\mathcal{A}_c^n$ be the nominal control law of the blocking controller and $\mathcal{Z}(0 \cdots \boldsymbol{K_e})$ the set of state sequences which are consistent with the specification in $\mathcal{N}_c$. Based on the decomposition procedure of (4.41), $\mathcal{A}_c^n \subseteq \mathcal{N}_c$, $\mathcal{A}_c^r \subseteq \mathcal{N}_c$, and $\mathcal{A}_c^n \neq \mathcal{A}_c^r$ hold. The latter implies that $\exists Z_c^r \in \mathcal{Z}(0 \cdots \boldsymbol{K_e})$, $Z_c^n \in \mathcal{Z}(0 \cdots \boldsymbol{K_e})$, and $Z_c^r \neq Z_c^n$ as different trajectories executed by $\mathcal{A}_c^r$ and $\mathcal{A}_c^n$, respectively. Thus, $\exists z_c^r \in Z_c^r$:

$$A^{|Z_c^r|}(z_c^r, z_{c0i}) > 0. \tag{5.21}$$

Similarly, $\exists z_c^n \in Z_c^n$:

$$A^{|Z_c^n|}(z_c^n, z_{c0i}) > 0. \tag{5.22}$$

Thus, $|A_w^{|Z_c^r|}(z_c^r, z_{c0i})| + |A_w^{|Z_c^n|}(z_c^n, z_{c0i})| > 1$. Since actuator faults are considered here, $|A_w^{|Z_c^r|}(z_c^r, z_{c0i}) \cup A_w^{|Z_c^n|}(z_c^n, z_{c0i})| \geq 2$ because $\mathcal{A}_c^r \neq \mathcal{A}_c^n$. Equation (5.16) implies that $\exists z_c \in \mathcal{Z}_c : rd_W(z_c) \geq 1$ so that (5.18) leads to $RD(\mathcal{N}_c) \geq 1 > 0$.

The safe feasibility of S in $\mathcal{N}_p^f$ results from the fact that the existence of $\mathcal{A}_c^r$ is equivalent with the existence of $\mathcal{N}_c$ because the latter is a supergraph of the former. According to Theorem 4.4, the existence of $\mathcal{N}_c$ implies the safe feasibility of S.

($\Longleftarrow$) Assume $RD(\mathcal{N}_c) = 0 \Rightarrow \exists \mathcal{A}_c^r$ for the faulty plant $\mathcal{N}_p^f$ and the specification S.

$$
\begin{aligned}
RD(\mathcal{N}_c) = 0, \quad &\Leftrightarrow \quad \max_{z_c \in \mathcal{Z}_c} \{rd_W(z_c)\} = 0 \\
&\Leftrightarrow \quad rd_W(z_c) = 0 \; \forall z_c \in \mathcal{Z}_c \\
&\Leftrightarrow \quad \left| \bigcup_{k=1}^{|\mathcal{Z}_c|-1} \boldsymbol{A}_w^k(z_{0c}, z_c) \right| = 1 \; \forall z_c \in \mathcal{Z}_c \\
&\Leftrightarrow \quad \exists k \in [1, |\mathcal{Z}_c| - 1] : |\boldsymbol{A}_w^k(z_{0c}, z_c)| = 1 \forall \; z_c \in \mathcal{Z}_c. \quad (5.23)
\end{aligned}
$$

The actuator failure means $\exists z_c^f \in \mathcal{Z}_c : \boldsymbol{A}_w^k(z_{0c}, z_c^f) = \emptyset$. The latter implies that $|\boldsymbol{A}_w^k(z_{0c}, z_c^f)| = 0$ which contradicts (5.23). Since the feasibility of S in $\mathcal{N}_p^f$ is linked with the condition $RD(\mathcal{N}_c)$ by an "and" conjunction, the invalid character of one of these conditions suffices to conclude the proof. $\qquad \square$

Note that Theorem 5.1 clearly states a need of redundancies in the controller in order to fulfill the specification for actuator failures. If a physical redundancy such as the alternative valve V_3 in the fluid level control process of Section 1.3.1 is considered, $RD(\mathcal{N}_c)$ will be greater than zero. However, the condition $RD(\mathcal{N}_c) > 0$ is necessary but not sufficient because, it does not offer any guarantee that the available redundancies are useful for the current reconfiguration problem. This guarantee is given by the safe feasibility condition.

Theorem 5.2 (Reconfigurability for actuator faults). *For a specification S, a control law $\mathcal{A}_c^{(1)} \subseteq \mathcal{N}_c$ and a faulty plant automaton $\mathcal{N}_p^f$ subject to an actuator fault, there exists a reconfigured controller $\mathcal{A}_c^r$ iff*

1. $\forall (z_s, w_s) \in \mathcal{Z}_s^f \times \mathcal{W}_s^f, |\mathcal{V}_{as}^f(z_s, w_s)| = 1$

2. S is safely feasible in $\mathcal{N}_p^f$ as described in Definition 4.7.

Proof of Theorem 5.2. ($\Longrightarrow$) The proof that the existence of $\mathcal{A}_c^r$ implies the safe feasibility of S in $\mathcal{N}_p^f$ is already proven in the previous Theorem 5.1 and is not repeated here. Hence, the implication is explained now for the first condition of Theorem 5.2.

Assume $\exists (z_s, w_s) \in \mathcal{Z}_s^f \times \mathcal{W}_s^f, : |\mathcal{V}_{as}^f(z_s, w_s)| \neq 1$. The case where $\mathcal{V}_{as}^f(z_s, w_s)| = 0$ is trivial because it directly confirms the assumption. Instead, the case $\mathcal{V}_{as}^f(z_s, w_s)| > 1$ is of interest. Observe that $\mathcal{V}_{as}^f(z_s, w_s)| > 1$ means that the supercontroller $\mathcal{N}_c$ is no longer W-deterministic w.r.t. Theorem 4.3 and violates the controllability condition of Theorem 4.4. If $\mathcal{N}_c$ is not W-deterministic and violates the controllability condition, then all sub-automata, including $\mathcal{A}_c^r$, will have the same properties as $\mathcal{N}_c$. Thus, the initial assumption lead to the conclusion that $\mathcal{A}_c^r$ is not W-deterministic and violates the reconfigurability, which means that it is invalid as a reconfigured controller.

($\Longleftarrow$) Assume that $\mathcal{A}_c^r$ does not exists, i.e., $\mathcal{A}_c = \emptyset$, but $\forall (z_s, w_s) \in \mathcal{Z}_s^f \times \mathcal{W}_s^f, |\mathcal{V}_{as}^f(z_s, w_s)| = 1$ and $\mathcal{S}$ is safely feasible in $\mathcal{N}_p^f$. $\mathcal{A}_c = \emptyset$ implies that $L_c^r(z_p', w_p, z_p, v_p) = 0$ $\forall (z_p', w_p, z_p, v_p) \in \mathcal{Z}_p^f \times \mathcal{W}_p \times \mathcal{Z}_p \times \mathcal{V}_p$. Hence, for any given couple $(z_s, w_s) \in \mathcal{Z}_s^f \times \mathcal{W}_s^f$, $|\mathcal{V}_{as}^f(z_s, w_s)| = 0$ holds and contradicts the first condition of Theorem 5.2. This contradiction makes the initial assumption invalid and concludes the proof. $\qquad\square$

The safe feasibility of the specification is required in Theorem 5.2 for actuator faults as in Theorem 5.1 for actuator failures. The difference between the two is that the redundancies are not explicitly required in the former as in the latter. Instead, W-determinism is required for the controller for actuator faults. This condition is necessary to insure that the reconfig-ured controller does not bring nondeterminism in the control after the reconfiguration step of (5.12).

Reconfigurability for system internal faults. Theorem 5.1 also hold for internal states faults where the plant deviates from its specified trajectory even though all actuators are healthy. However, it is possible, in certain cases, to model a system internal fault as if the correctly activated actuator is malfunctioning. In this way, the reconfigurability problem for system internal faults is transformed into a reconfigurability problem for actuator faults or failures, where Theorem 5.1 and Theorem 5.2 can be applied.

Contrary to actuator failures and internal faults, redundancies are not always required for reconfiguration after a sensor fault. The feasibility of the specification $\mathcal{S}$ in $\mathcal{N}_p^f$ is sufficient to guarantee reconfigurability.

Theorem 5.3 (Reconfigurability for sensor faults). *For a specification $\mathcal{S}$, a control law $\mathcal{A}_c^{(i)} \subseteq \mathcal{N}_c$ and a faulty plant automaton $\mathcal{N}_p^f$ subject to a sensor fault, there exists a reconfigured controller $\mathcal{A}_c^r$ iff $\mathcal{S}$ is safely feasible in $\mathcal{N}_p^f$ as described in Definition 4.7.*

Proof of Theorem 5.3. ($\Longrightarrow$) This direction of the proof is almost trivial. $\exists \mathcal{A}_c^r \Rightarrow \mathcal{N}_c^r \neq \emptyset$. Hence, Theorem 4.4 implies that $\mathcal{S}$ be safely feasible.

($\Longleftarrow$) Assume $\mathcal{S}$ is not feasible and $\exists \mathcal{A}_c^r$ for $\mathcal{S}$ in $\mathcal{N}_p^f$. Let $\mathcal{A}_c^n$ be the nominal blocking controller. Since all three specifications $\mathcal{S} \models Z_s$, $\mathcal{S} \models z_F$, and $\mathcal{S} \models W_s$, can be expressed as set of state sequences $Z_s(0 \cdots K_e)$, a state sequence is used in this proof. Consider the specification $\mathcal{S} \models Z_s$ as being not safely feasible which means that Z_s is either not basically feasible or the plant can deviate from Z_s due to its nondeterminism (cf. Theorem 4.1). The case of the basic feasibility is trivial whereas the deviation from Z_s requires some more investigation. In case of a deviation, (4.28) leads to

$$\bigvee_{k=0}^{k_e-1} \bigvee_{z_p'}^{Z_p \backslash z_{sk}'} \bigvee_{w_p}^{W_p} \bigvee_{v_s}^{V_{ap}(z_{sk}', z_{sk})} L_p^f(z_p', w_p, z_{sk}, v_s) = 1 \tag{5.24}$$

with $z_{sk} = Z_s(k)$ and $z_{sk}' = Z_s(k+1)$. Equation (5.24) means that $\exists k \in (0, k_e - 1]$: $Z_p(k+1) \neq Z_s(k+1)$ with $v_s \in V_{ap}(z_{sk}', z_{sk})$. Since sensor faults are considered here, (5.13) will be the selected reconfiguration method. This method will adapt the input of $\mathcal{A}_c^r$ to the wrong sensor values but let the output unchanged. Thus, $\mathcal{A}_c^r$ will share the same output events v_s as the nominal blocking controller $\mathcal{A}_c^n$. Hence, $\mathcal{A}_c^r$ will still violate the specification as described by (5.24). Thus, $\mathcal{A}_c^r$ is invalid as a reconfigured controller. The fact that the initial assumption is wrong concludes the proof. $\square$

5.5. Complexity analysis of the reconfiguration

The complexity of the on-line reconfiguration algorithm proposed in this thesis is now analyzed regarding the required time of execution. The complexity of the reconfiguration algorithm (Algorithm 3) is investigated here to check the tractability of the reconfiguration problem. Only expressions which may need more than one execution step are discussed in the following.

Complexity of the strategy selection: $\mathcal{C}_{31}$. Line 2 will need $|\mathcal{V}_p^f| \cdot |\mathcal{W}_p^f|$ execution steps in the worst case. The signal sets of the faulty plant may differ from those of the nominal plant only by the empty symbol ε which characterizes failures. Thus, $|\mathcal{V}_p^f| \cdot |\mathcal{W}_p^f| = (|\mathcal{V}_p| + 1) \cdot (|\mathcal{W}_p|+1)$. The preparation of forward recovery in Lines 5 and 6 needs $|\mathcal{V}_c| \cdot |\mathcal{W}_c| + |\mathcal{V}_c| \cdot |\mathcal{W}_c|$ steps. In the worst case, the cardinality of the signal sets of the controller is the same as for the plant. Thus, the required number of executions for the forward reconfiguration preparation is at worst $2|\mathcal{V}_p||\mathcal{W}_p|$. Therefore, the complexity of the strategy selection is

$$\mathcal{C}_{31} = 1 + (|\mathcal{V}_p| + 1)(|\mathcal{W}_p| + 1) + 1 + 2|\mathcal{V}_p||\mathcal{W}_p|. \tag{5.25}$$

Complexity of the computation of $\overline{Z}_s(k_{sf} \ldots k_{sf} + l)$: $\mathcal{C}_{32}$. The *while* condition in Line 8 checks if an input signal belongs to the active inputs of two states. The state $\overline{Z}_{\mathrm{BFS}}(1)$ does not change because it is the blocking one whereas $\overline{Z}_{\mathrm{BFS}}(2)$ changes depending on the Breadth-First-Search (BFS) result. In the worst case the whole state space must be searched until the BFS result is nonempty, i.e, in at most $|\mathcal{Z}_p| - 1$ steps. The maximal number of acive inputs between the considered states is $|\mathcal{V}_c| \leq |\mathcal{V}_p|$. In addition, each evaluation of the second condition in Line 8 consists of an addition and a comparison. Hence, maximal number of execution of the *while* condition is $(|\mathcal{Z}_p| - 1) \cdot |\mathcal{V}_p| \cdot 2$.

The first part of the *if* condition in Line 9 consists of at most $|\mathcal{Z}_p| - 2$ steps in the worst case so that at least the start state and the target state are connected. The second part of the *if* condition Line 9 consists of one comparison and the same condition as the first part of the *while* condition. Thus, the complexity of this *if* closure from Line 9 to Line 12 is $(|\mathcal{Z}_p| - 2) + (|\mathcal{Z}_p| - 1) \cdot |\mathcal{V}_p| + 3$.

The complexity of the *else* closure from Line 13 to Line 19 solely depends on the complexity of the BFS. In a graph made of E edges and V vertices, the BFS is known to have a complexity of $\mathcal{O}(V + E)$ [86]. In the context of nondeterministic automata, the complexity of the BFS is deduced to be one comparison added to $|\mathcal{Z}_p| + |\mathcal{V}_p||\mathcal{W}_p|$ for the worst case. Therefore, the complexity of the computation of $\overline{Z}_s(k_{sf} \ldots k_{sf} + l)$ can now be summarized to

$$\mathcal{C}_{32} = 2|\mathcal{V}_p|(|\mathcal{Z}_p| - 1)((|\mathcal{Z}_p| - 2) + (|\mathcal{Z}_p| - 1) \cdot |\mathcal{V}_p| + 3 + |\mathcal{Z}_p| + |\mathcal{V}_p||\mathcal{W}_p| + 1) + 1. \quad (5.26)$$

Complexity of the computation of L_{c-set1}: $\mathcal{C}_{33}$. The number of iterations in the *for* loop depends on the length of the BFS result. In a nondeterministic automaton $\mathcal{N}_p$ there are at most $|\mathcal{Z}_p| - 1$ iterations. Line 24 needs at worst $|\mathcal{V}_c| + |\mathcal{W}_c|$ executions. Since the controller has at most as many input and output events as the plant, Line 24 needs at most $|\mathcal{V}_p| + |\mathcal{W}_p|$ steps. The remaining expressions of this *for* loop need one execution step each. Thus, the computation of the set of transitions for which the characteristic function of the reconfigured controller should be set to 1 (Line 22 to Line 28) leads to:

$$\mathcal{C}_{33} = (|\mathcal{Z}_p| - 1)(2 + |\mathcal{V}_p| + |\mathcal{W}_p| + 2). \quad (5.27)$$

The complexity of the complete algorithm is now given by

$$\mathcal{C}_3(n) = \mathcal{C}_{31} + \mathcal{C}_{32} + \mathcal{C}_{33}. \quad (5.28)$$

The choice of the variable n is based on the fact that the complexity of an automaton graph depends on the number of vertices and edges. They are determined here by $|\mathcal{Z}_p|$, $|\mathcal{V}_p|$ and $|\mathcal{W}_p|$. Since the size of each symbol set is not predictable, it is assumed that they all have the same size

$$n = |\mathcal{Z}_p| = |\mathcal{V}_p| = |\mathcal{W}_p|. \tag{5.29}$$

By applying (5.29) on (5.25), (5.26), (5.27) in (5.28), the complexity of Algorithm 3 is obtained as follows:

$$
\begin{aligned}
\mathcal{C}_3(n) &= 1 + (n+1)(n+1) + 1 + 2n^2 + 2n(n-1)[(n-2) + n(n-1) + \\
&\quad 3 + n + n^2 + 1] + 1 + (n-1)(2 + n + n + 2) \tag{5.30} \\
&= 4n^4 + 2n^3 + 3n^2 - 2n + 2 \tag{5.31} \\
\Rightarrow \mathcal{C}_3(n) &\in \mathcal{O}(n^4). \tag{5.32}
\end{aligned}
$$

Equation (5.32) shows that the on-line reconfiguration problem is solved by Algorithm 3 in polynomial time. Therefore, the reconfiguration problem can be said to be tractable.

5.6. Control reconfiguration of the fluid process

The aim of this section is to explain how to apply the control reconfiguration method presented in this thesis on the fluid level control process of Section 1.3.1. Compared to the other examples of Chapter 6 and Chapter 7, it has the advantage that the reconfiguration steps can be manually followed. The off-line control reconfiguration algorithm is then applied in a simulation for the case of a blocking valve.

5.6.1. Faulty behavior after an actuator failure: broken outflow valve

The actuator fault considered here is the blocking valve V_2 in Fig. 1.2 for which the FDI unit generates the fault event $f = v^f = 2$. Assume that the fault is detected at step $k_f = 5$ with $z_p(k_f) = 4$. The effect of the fault on the plant automaton is described by the self-loop $2/4$ at state 4 in Fig. 3.5.

Figure 5.2 demonstrates the simulation of the nominal controller $\mathcal{A}_c^{(1)}$ with the faulty plant $\mathcal{N}_p^f$. The level of Tank T_1 remains at level 4 despite the "open valve V_2" command from the controller. In fact the controller automaton blocks because it does not expect the output 4 from state 4 but the output 3 instead. Hence, the command $w_c = 2$ which means to "open valve V_2 and close the others" is executed only by the pump P_1 and the valve V_3. No reaction is registered from valve V_2 so that $w_c = 2$ has the same effect on the plant as $w_c = 0$. Thus, there exists an input error function $E_v(\cdot)$ for which $E_v(2) = 0$ holds.

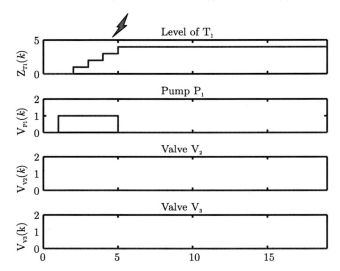

Figure 5.2.: Simulation of the behavior of the faulty level control process

5.6.2. Application of the off-line control reconfiguration

Reconfigurability. The reconfigurability condition for this actuator failure is now investigated w.r.t. Theorem 5.1:

1. Let $Z_c(0 \cdots 5) = (0, 1, 2, 3, 4, 3)$ be the state sequence executed so far in the supercontroller $\mathcal{N}_c$ (Fig. 4.14).

2. Equation (5.15) leads to $rd_W(3) = 2$ for the considered state sequence Z_c. Hence, (5.18) implies that $RD(\mathcal{N}_c) > 0$. The first condition of Theorem 5.1 is fulfilled.

3. Consider the faulty plant $\mathcal{N}_p^f$ of Fig. 3.5. Equation (4.28) yields zero since there is no risk of deviation from the specification. The second condition of Theorem 5.1 is therefore fulfilled.

Therefore, there exists a reconfigured controller $\mathcal{A}_c^r$.

Off-line reconfiguration steps. The off-line reconfiguration steps of Section 5.1.2 are now demonstrated:

1. The model of the faulty plant $\mathcal{N}_p^f$ is depicted in Fig. 3.5

2. The blocking control policy with its trajectory are read:

$$\mathcal{A}_c^r = \mathcal{A}_c^{(1)} \tag{5.33}$$

$$\overline{Z}_c^r = \overline{Z}_c^{(1)} \tag{5.34}$$

3. Transitions where the broken actuator is required are deleted with $v_p^n = 2$. Thus,

$$L_c^r(z', 2, z, w) = 0, \forall (z', z, w) \in \mathcal{Z}_p^2 \times \mathcal{W}_p \tag{5.35}$$

$$\tag{5.36}$$

4. In this example, ε could be defined as to deactivate all actuators.

5. Synchronize the states of $\overline{Z}_c^r$ with $\mathcal{N}_p^f$

$$\overline{Z}_c^r(k_i) = \{4\}. \tag{5.37}$$

6. The next healthy state is 3, thus $l = 1$ with (5.8).

7. The iterative computation of $\mathcal{Z}^r(k_i \cdots k_i + 1)$ leads to the sequence $(4, 3)$

8. There is only one sequence in $\mathcal{Z}^r(k_i \cdots k_i + 1)$, hence, the state sequence selection here is trivial.

9. The only transition available is the one labeled with $3/3$ from state $z_p^f = 4$. Equation (5.10) leads to $L_c^r(3, 3, 4, 3) = 1$.

The resulting fault-tolerant controller $\mathcal{A}_c^r$ is depicted in Fig. 5.3. The simulation of the

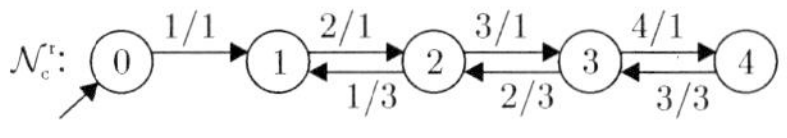

Figure 5.3.: Fault-tolerant controller of the faulty level control process

fault-tolerant controller is depicted in Fig. 5.4. After the fault has been detected, the reconfiguration procedure runs between step 5 and 6. The simulation result confirms the heuristical expectations since the valve V_3 is now used to achieve the specification (4.26).

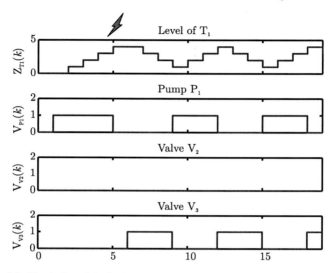

Figure 5.4.: Simulation of the faulty level control process with a fault-tolerant controller

5.6.3. Discussion

The reconfiguration step of the controller for this example consists of deleting the faulty transitions and adding the redundant ones in the control policy automaton. However it confirms the intuitiv expectations by means of a heuristic approach. Remark in Fig. 1.2 that there is no redundancy for the pump P_1. Hence, for the fault "stuck at closed pump P_1" even a controller redesign would fail because $\mathcal{N}_s^f = \emptyset$ would hold by applying (4.17) with (4.24) on $\mathcal{N}_p^f$. Note that the test automaton would be empty and the specification would not be even basically feasible w.r.t Definition 4.6.

This shows how a discrete-event control reconfiguration may fail despite the presence of physical redundancies for which $RD(\mathcal{N}_c) > 0$ even hold. In contrast to the previous fault scenario, the redundant valve V_3 is useless for this failure.

Part II.

Application examples

6. Virtual manufacturing process: Pick-and-place system

Abstract. *This chapter presents a virtual pick-and-place system which is used to highlight the use of the on-line control reconfiguration algorithm. A particular emphasis is put the demonstration of the forward and backward control reconfiguration after a system internal fault and a sensor fault respectively.*

6.1. Modeling and nominal control design of the pick-and-place system

The pick-and-place system (PPS) introduced in Section 1.3.1 is modeled by an nondeterministic I/O automaton $\mathcal{N}_p$. Each transition is labeled with the I/Os v_p/w_p. Each state z_p is represented by three elements standing each for the number of workpieces placed in the corresponding column (see Fig. 6.1). Hence, $z_p = 320$ reveals that the 1st column has 3 workpieces, the 2nd column 2 workpieces and the 3rd is empty. The transport of a workpiece from B_1 to the correct hole in the box of B_2 is labeled by the command $v_p = 1$. The command $v_p = 2$ takes the box out of PPS, whereas $v_p = $ x is used to take a workpiece from the box to the trash. Possible outputs are the measurements $w_p = 1, 2$ or 3 and $w_p = 10, 20$ or 30 from the visual sensor *VS* and the thermal sensor TS, respectively. The transport of a workpiece with the ID 1, 2 or 3 leads to an incrementation of the number of workpieces in the corresponding column of a given state z_p.

Figure 6.2 is a compact representation of the possible transitions from every state out of the set $\mathcal{Z}_p = \{000, \ldots, 333\}$ which has 64 elements presented in Fig. 6.3. Instead, only the relevant part which is needed to demonstrate the approach is depicted in Fig. 6.1. The incrementation of the number of workpieces in a column of the box results into an incrementation of the 1st, 2nd or 3rd variable i, j or k as depicted in Fig. 6.2 through the $^+$ symbol. The approach is applicable to the whole plant automaton since every state has similar transitions as the state ijk in Fig. 6.2.

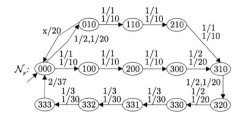

Figure 6.1.: Relevant part model of the PPS

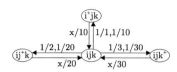

Figure 6.2.: Modeling procedure of the PPS

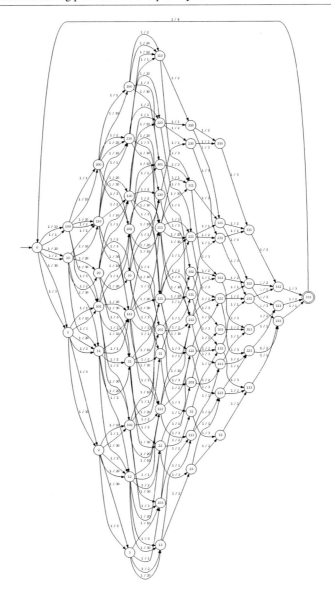

Figure 6.3.: Overview of the complete I/O automaton of the PPS

For this manufacturing process, the specification S is expressed as the final state $z_F = 333$ which shows that all the three column are full with the right workpieces inside. A redundance-free control law $\mathcal{A}_c$ fulfilling this specification is obtained as explained in Section 4.3 and depicted in Fig. 6.4.

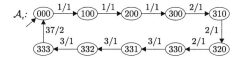

Figure 6.4.: Nominal controller of the PPS

6.2. Forward control reconfiguration after an internal system fault: random workpiece supply

The internal system fault f_1 described in Section 1.3.1 leads to the situation where the controller does not transport the "wrong" workpieces to the box but let them fall into the trash. Assume that the box is empty ($z_p = 000$). When f_1 occurs, the controller remains in state $Z_c(k_{sf}) = 000$ and performs a self-loop labeled with $2/2$ and $3/2$ until a workpiece with the ID 1 enters the system (Fig. 6.5). These self-loops symbolize the repairing strategy whereas the control output of the previous step $w_c(k_{sf} - 1) = 2$ remains unchanged. According to the model it means that the controller now tries to take the new empty box out of the PPS. However the command $w_c(k_{sf}) = 2$ can not be executed by the plant which is in state $z_p = 000$. As a result the control loop is in a blocking situation. Assume that a workpiece with the ID 2 is the next one at step k_{sf}, the on-line reconfiguration algorithm works as follows.

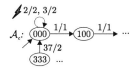

Figure 6.5.: Blocking controller due to random workpieces supply

Strategy selection. Since the fault is diagnosed at the moment where the controller blocks, $k_{sf} = k_s$ holds. Hence, a *forward recovery* has to be prepared by building L_{c-set0}:

$L_c(\overline{Z}_s(k_{sf}+1), *, \overline{Z}_s(k_{sf}), \mathcal{V}_{ac}(\overline{Z}_s(k_{sf}+1), \overline{Z}_s(k_{sf}))) = L_{c-set0}(100, 1, 000, 1) = 0$ which means that the transition from state 000 to state 100 labeled $1/1$ is disabled.

Computation of $\overline{Z}_s(k_s \ldots k_s+l)$. A Breadth-First-Search is performed in $\mathcal{N}_p^f$ from the current faulty state $\overline{Z}_s(k_{sf})$ to each state of the remaining state sequence $\overline{Z}_s$ by incrementing the counter n until a reachable state is found. This is expressed by $\mathrm{BFS}(\overline{Z}_s(k_{sf}), \overline{Z}_s(k_{sf} + n), \mathcal{N}_p^f)$ in Algorithm 3. The next state in $\overline{Z}_s$ which can be used for a workpiece with the ID 2 is the state 310. Therefore, the $\mathrm{BFS}(\cdot)$ operator returns the sequence $\overline{Z}_{\mathrm{BFS}} = (000, 010, 110, 210, 310)$. The value l is then obtained w.r.t. (5.20) as $l = n = 3$ and $\overline{Z}_s(k_{sf} \ldots k_{sf} + 3)$ replaces corresponding states in $\overline{Z}_s(0 \cdots k_e)$.

Build L_{c-set1} **for** $\mathcal{A}_c^r$. Now the first transitions from which the characteristic function should have the value 1 are derived from $\overline{Z}_s(k_{sf} \ldots k_{sf} + 3)$. In this case $L_{c-set1} = \{(010, 1, 000, 2); (110, 1, 010, 1); (210, 1, 110, 10); (310, 1, 210, 1); (310, 1, 210, 10)\}$ is added to the control law $\mathcal{A}_c$ to obtain the reconfigured control law $\mathcal{A}_c^r$ (Fig. 6.6). The first tuple of L_{c-set1} is used by the reconfigured controller to transport the wrong workpiece with the ID 2 through the command $w_c(k_{sf} + 1) = 1$ to its right column in the box. Afterwards, the other transitions are useful only in the case that triangular workpieces enter the process in the expected order.

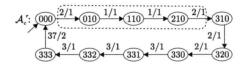

Figure 6.6.: Reconfigured transitions due to random workpieces supply

6.3. Backward reconfiguration after a sensor fault: damaged visual sensor

In this case the visual sensor VS is malfunctioning and always returns the value 2 because it sees every workpiece as having the ID 2. Assume that the plant is at state 000 while a workpiece with the ID 1 arrives in front of VS, before the fault occurs. The considered workpiece is then wrongly read as having the ID 2. Consequently, the workpiece is transported to the second column of the box, hence, the state 010 is reached by the plant whereas the controller remains at state 000 because the expected plant output $w_c(k) = 1$ is not coming.

Strategy selection. The diagnosis unit identifies the occurrence of the fault at step k_{sf} which occurred in the past at state $Z_p(k_s - 1) = 000$ as shown in Fig. 6.7. Hence, $k_{sf} < k_s$ holds and makes a *backward recovery* more appropriate in this case. According to the reconfiguration algorithm, the correction term k_c is set to -1 to implement the backward recovery.

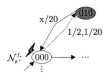

Figure 6.7.: Faulty plants transition due to a sensor fault

Computation of $\overline{Z}_s(k_s \ldots k_s + l)$**.** The BFS$(\cdot)$ operator is used as in the previous case whereas $z_{start} = Z_p(k_{sf} + 1) = 010$ is the actual state of the faulty plant and $z_{end} = \overline{Z}_s(k_{sf}) = 000$ the state of the blocking controller. In this example only one round is sufficient to find a state transition from 010 back to 000 as $\overline{Z}_{\mathrm{BFS}} = (010, 000)$. The next step is to insert this result of the Breadth-First-Search into $\overline{Z}_s(0 \cdots k_e)$ without overwriting healthy transitions. This results in replacing the state 000 through 010 and shifting the other states in $\overline{Z}_s$. Thus, $l = 2 + 9 - 2 \cdot 0 - 1 = 10$ w.r.t. (5.20). As a result, $\overline{Z}_s(k_s \ldots k_s + l) = (010, 000, 100, \ldots, 333)$.

Build L_{c-set1} **for** $\mathcal{A}_c^r$**.** It only consists of the transition

$$(000, \mathcal{V}_{ap}^f(000, 010), 010, \mathcal{W}_{ap}^f(000, 010)) = (000, \mathrm{x}, 010, 20).$$

It means that the controller should drive the plant back to the healthy state 000 by taking the wrong workpiece with the ID 2 to the trash.

Discussion. The solution of throwing the wrong parts in trash might be economically not the best but would keep the process running. A better solution would be to correctly sort the workpieces without throwing them away. However this leads to a more complex model involving events enabling the transport of workpieces from a wrong hole to a correct one by using the thermal sensor TS.

7. Experimental manufacturing process

Abstract. *This chapter describes the applicability of fault-tolerant control method of this thesis in an industrial environment. Therefore, a real manufacturing cell is considered as the plant for which the signals sent and received by the PLC need to be adapted to the faults. Since this plant has a higher complexity than those presented in previous chapters, a component-oriented modeling of the manufacturing cell into an I/O automata network is applied. A central controller is then derived for the I/O automata network. Experimental results of the on-line control reconfiguration algorithm are presented for actuator and system internal faults.*

In this section the applicability of the reconfiguration method proposed in Section 5.3 is demonstrated on a real plant by means of physical experiments. The experiments were performed on the pilot manufacturing cell from the Institute of Automation and Computer Control at the Ruhr-Universität Bochum. Figure 7.1 shows a photograph of the plant.

7.1. Structure of the manufacturing cell

Physical description of the pilot plant. A structural overview is given in Fig. 7.2. The pilot plant is divided in two zones of activity. From the dashed line in Fig. 7.2, the right domain is the reserved for the horizontal gripper and the left domain is dedicated to the vertical gripper.

The horizontal gripper domain consists of the following components:

- Horizontal handler (HH)

- Horizontal handling gripper (GHH)

- 21 positions in the following blocks:

  - Heating block (HB) with 4 positions

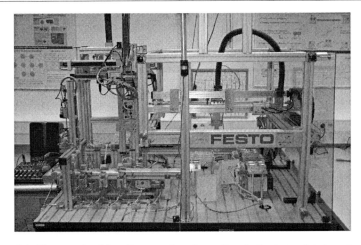

Figure 7.1.: Photograph of the pilot manufacturing cell at the Institute of Automation and Computer Control at the Ruhr-Universität Bochum

- Cooling block (CB) with 4 positions

- Measuring block (MB) with 4 positions

- Deposit (D) with 9 positions.

Vertical gripper domain consists of the following components:

- Vertical handler (VH)

- Vertical handling gripper (VHG)

- Deflectors (D1,...,D4)

- Slides (Sl1,...,Sl4)

- Magazine (M)

- Pusher (P).

Components which both belong to the left and right domain are the following:

- Buffer rail (BR)

- Conveyor belt (B)

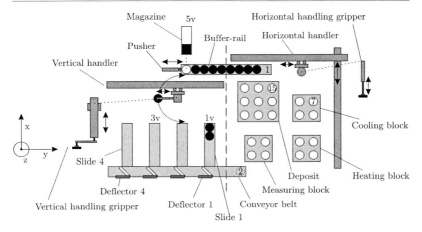

Figure 7.2.: Schematic of the overview of the pilot manufacturing cell

Moreover, the manufacturing cell consists of five extra deposit places and a station for control of quality. Since these components are not relevant for the considered process, they are not modeled here. The modeling of the components enumerated above is explained in Section 7.2.

Interaction with the PLC and MATLAB/Simulink. Figure 7.3 shows how the experimental setup is divided into three classic levels of process automation, i.e, the supervision level, the execution level and the process level. The discrete-event supervision of the manufacturing cell is implemented in MATLAB/Simulink. The following steps are achieved in MATLAB:

- Modeling of the plant as a network of extended I/O automata

- Definition of the specification of the process

- Controller design as an I/O automaton

- Setup of simulation parameters.

The following steps are achieved in Simulink:

- Exchange of commands and measurements with the PLC

- Controller realization

- Diagnosis

- Control reconfiguration.

Although the last three items run in a Simulink environment, they are implemented as Embedded MATLAB Functions. Actuator and sensor faults are activated in Simulink whereas process fault provoked through external means acting directly on the plant.

Once the process is running, the Simulink model exchanges arrays of inputs $[v_1 \ldots v_8]$ and outputs $[w_1 \ldots w_{12}]$ with a PLC of type Siemens S7-300 in UDP-datagrams over an Ethernet connection. One role of the PLC is to collect sensor data from the digital input cards, convert them into output events from the plant to be processed by the controller in Simulink (see Fig. 7.4). According to the specification, the controller generates output events in Simulink which are sent to the PLC (see Fig. 7.5). At the execution level, the PLC sets its process image output register according to the output events coming from the Simulink model. The measurements from the plant are written on the process image input register of the PLC and sent up to the Simulink model.

The signal converter connects the execution level to the process level. This interface converts the PLC output signals into adequate voltage to activate actuators, e.g., the conveyor belt, the vacuum of the grippers. The conversion of physical information into PLC input signals is also handled by this interface.

7.2. Component-oriented modeling of the manufacturing cell

In order to model the behavior of the manufacturing cell is divided in several components which are first modeled as extended I/O automata. Based on the interdependencies among the components, the coupling matrix is built and connected with the extended I/O automata to obtain the I/O automata network depicted in Fig. 7.6. The relevant aspects considered during the modeling procedure are now presented for each component of the manufacturing cell.

Horizontal handler (HH). The horizontal handler is able to move the horizontal handling gripper among the buffer rail, the deposit and the conveyor belt. Hence, the control input v can only be a target position to reach. The next states z' and output w are the current position. The horizontal handler influences the behavior of the horizontal gripper over the interconnection output signal. In addition, this handler "notifies" its presence over the other components that it can reach in the plant. This notification occurs through the second interconnection output signal. For safety reasons the horizontal handler is not allowed to move if its gripper is in certain positions, e.g., down holding a workpiece. The authorization to move comes from the horizontal gripper as an interconnection input signal to horizontal handler. The model of the horizontal handler is given in Section D.4.

Horizontal and vertical handling gripper (GHH and GVH). Each gripper makes use of a vacuum to pick-up workpieces, to hold them or drop them in any position the horizontal or vertical handler has reached. Thus, the control input switches the vacuum on and off or move the gripper up an down. The control output tells the position of the gripper and the presence or absence of a workpiece thereunder. The picking-up and dropping events of workpieces from any position are exchanged through the second interconnection output. The model of the horizontal handling gripper is given in Section D.5 and the vertical handling gripper in Section D.11.

Positions in blocks (Pos). Each of 21 positions in the heating block, the cooling block, the measuring block and the deposit is considered as a component without control input and output since they can only be influenced by internal means, i.e, workpieces. Moreover, there is no sensor enabling to check if a position is occupied or not. This is the reason why their model only have interconnection input signals. The model of the positions in blocks is given in Section D.6.

Vertical handler (VH). The vertical handler is able to move the vertical hanlding gripper among the slides and the magazine. The control input and output events as well as the interconnection signals are similar to the horizontal handler model. Intermediary movements of the pivoting arm between the slides and the magazine are handled by low-level controllers. The model of the vertical handler is given in Section D.10.

Deflectors (D1,...,D4). The role of this components is to direct workpieces from the conveyor belt into the corresponding slides. If activated, they move from their straight position to the oblique position so that the workpieces are forced to slip into the slides. Thus, the presence of every new workpiece notified to the corresponding slide model through the interconnection output signal whereas the state the other deflectors is notified by the conveyor belt model through the interconnection input signal. The model of the vertical handler is given in Section D.8.

Slides (S1,...,S4). The slides receive the workpieces directed to them by the corresponding deflector. A maximum of 6 workpieces can be aligned in a slide. The vertical gripper can take out those workpieces but can not drop them back because the other workpieces, if any, would slip down to the bottom of the slide. Sensors placed at the top and at the bottom of each slide detect if it is empty, full or with an intermediary number of workpieces in between. Since they can not be directly controlled, they have no control input signals. The model of the slides is given in Section D.9.

Magazine (M). The magazine can store up to 10 workpieces as a stack. The workpieces can be added at the top by the vertical gripper and removed by the pusher at the bottom. This internal influence is reflected by the interconnection input signals and explains the absence of external control input signals. Sensors placed at the top and the bottom of the magazine detect if it is empty, full or with an intermediary number of workpieces in between. This sensor information is sent out through control output signals. The model of the magazine is given in Section D.3.

Pusher (P). This component pushes workpieces from the bottom of the magazine into the buffer rail. When the pusher is pulled back, the next workpiece, if any, falls down from the stack. In the case that the buffer rail is full, the pusher can not be completely extended. The model of the pusher is given in Section D.1.

Buffer rail (BR). The workpieces on the buffer rail are pushed from the bottom of the magazine by the pusher. Workpieces can be removed from the buffer rail only from its end by the horizontal gripper. However, they can not be drop back to that position. This internal interactions happens through the interconnection signals. Thus, there are no control input signals but a control output revealing if there is workpiece ready for transport at the end of the buffer rail. For the considered process, the buffer rail is always filled with enough workpieces so that there is no empty spaces between the workpieces. The model of the buffer rail is given in Section D.2.

Conveyor belt (B). The conveyor transports workpieces droped by the horizontal gripper along the deflectors and their corresponding slides from the first to the fourth. The position of the workpieces on the belt is registered by sensors placed next to the slides. That information is sent out as a control output which is equal to the next state the model of the belt is moving to. It is not possible to switch the direction in which the belt is moving. The model of the conveyor belt is given in Section D.7.

Complexity issues. Due to the state space explosion phenomenon, it is not possible to compute a composed I/O automaton out of the I/O automaton network of the manufacturing cell. Instead, the resulting I/O automata network is simulated with vectorial input sequences where each element of the input vector is foreseen for a specific component. The next states and control output of the I/O automata network are also vectors where each element reflects the behavior of the corresponding component. However, to simplify the interpretation and analysis of those sequences, each vector is encoded into a scalar symbol. The encoded simulation results of the I/O automata network are identical to those of corresponding composed I/O automaton but obtained with far less computational efforts [10]. This is justified by the fact that the composition algorithm is executed only one step. Therefore, this technique is titled as *on-line composition* and is less complex than the classic composition.

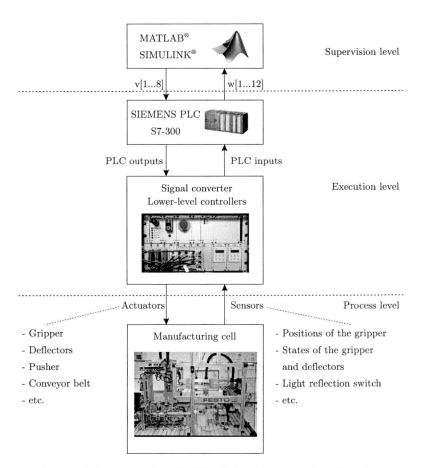

Figure 7.3.: Interaction of the FESTO with the PLC and MATLAB/Simulink

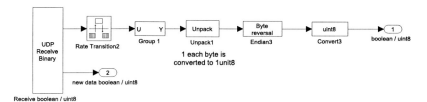

Figure 7.4.: Receiver Simulink block

Figure 7.5.: Sender Simulink block

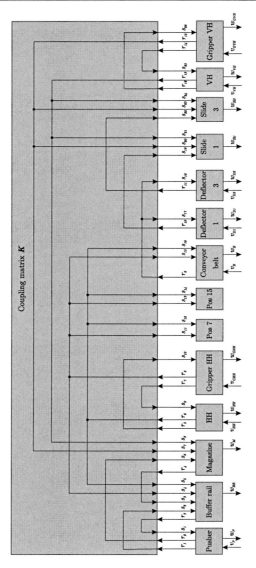

Figure 7.6.: I/O automata network of the pilot manufacturing cell

7.3. Nominal manufacturing process

The process used here to demonstrate the applicability of the developed methods in an industrial environment consists of transporting a workpiece through several stations of the plant. In an industrial environment, each station could be responsible for screwing, welding, cooling, cleaning workpieces or a product. The main steps to be achieved during the process are depicted in Fig. 7.7 as follows:

1. Place a workpiece on position 1 in the buffer rail to be picked-up by the horizontal gripper.

2. Move the workpiece from position 1 to position 15 in the deposit which represents the first station.

3. Move the workpiece from position 15 to position 7 in the cooling block which represents the second station.

4. Move the workpiece from position 7 to position 2 on the conveyor belt which represents the last station.

5. Activate the conveyor belt and deflector D1 in order to direct the workpiece to the bottom of slide Sl1.

6. Move the vertical gripper to the bottom of slide Sl1 to pick-up the workpiece and drop it into the magazine for the next round.

7. Go back to Step 1.

The verbal description of the process in Fig. 7.7 is now formalized as a state sequence specification $\mathcal{S} \models Z_s(0 \cdots k_e)$. The following matrix represents the specified vectorial sequence of the automata network shown in Fig. 7.6. Each column represents the state sequence of a single component. For instance, the fourth column of $\boldsymbol{Z_s}(0 \ldots 12)$ shows the required state sequence of positions for the horizontal handler as $Z_{s-HH}(0 \ldots 12) = (1, *, 1, *, 15, *, 7, *, 1, *, 1, *, 1)$. Each row $\boldsymbol{Z_s}(k)$ of $\boldsymbol{Z_s}(0 \ldots 12)$ is a vector that represents the required state for each component of the I/O automata network at step k. The $*$ symbols between the rows are unspecified states. They describe degrees of freedom among the states

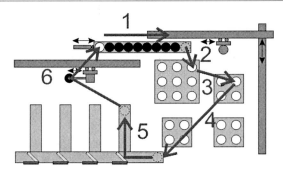

Figure 7.7.: Overview of the nominal process in the pilot manufacturing cell

of the specification. The process is a cycle because the first and the last row of $Z_s(0\ldots 12)$ are identical.

$Z_s(0\ldots 12) =$

k	P	BR	M	HH	GHH	Pos7	Pos15	B	D1	D3	S1	S3	VH	GVH
0	1	1	1	1	1	1	1	1	1	1	1	1	1	1
1	*	*	*	*	*	*	*	*	*	*	*	*	*	*
2	1	2	2	1	1	1	1	1	1	1	1	1	1	1
3	*	*	*	*	*	*	*	*	*	*	*	*	*	*
4	1	1	2	15	1	1	2	1	1	1	1	1	1	1
5	*	*	*	*	*	*	*	*	*	*	*	*	*	*
6	1	1	2	7	1	2	1	1	1	1	1	1	1	1
7	*	*	*	*	*	*	*	*	*	*	*	*	*	*
8	1	1	2	1	1	1	1	2	1	1	1	1	1	1
9	*	*	*	*	*	*	*	*	*	*	*	*	*	*
10	1	1	2	1	1	1	1	1	1	1	1	1	1	2
11	*	*	*	*	*	*	*	*	*	*	*	*	*	*
12	1	1	1	1	1	1	1	1	1	1	1	1	1	1

$$(7.1)$$

The resulting I/O automaton $\mathcal{A}_{c-main}$ representing the control law is depicted in Fig. 7.8. The states, inputs and outputs used in $\mathcal{A}_{c-main}$ are coded vectors into scalar values. The vectors consist of elements related each to a component of the I/O automata network. The complete encoding tables of the all states and signals used in $\mathcal{A}_{c-main}$ are shown in Section D.12. The internal state sequence of the controller is

$$
\begin{aligned}
Z_{c-main} =&(1, 89, 86, 87, 88, 52, 80, 84, 82, 72, 82, 84, 80, 64, 69, 67, 62, 67, 69, 64, 52, 55, \\
&59, 57, 53, 46, 47, 48, 45, 41, 42, 43, 33, 35, 36, 3, 2, 1).
\end{aligned} \tag{7.2}
$$

The control set sequence Z_c is an element of the set of state sequences represented by the specification defined in (7.1). The encoded state specification sequence is

$$
\mathbf{Z_s}(0 \dots 12) = (1, *, 86, *, 72, *, 62, *, 46, *, 33, *, 1)^T. \tag{7.3}
$$

Note that the intermediary states or degrees of freedom in the specification symbolized by the $*$ symbol in (7.3) are explicitly specified in the control sequence of (7.2).

The behavior of the closed-loop system with the controller $\mathcal{A}_{c-main}$ and the manufacturing plant is depicted in Fig. 7.9. This plot shows an overview of the input, output and state sequences recorded during the nominal process. The first plot of Fig. 7.9 illustrates the input sequence $\tilde{v}_p = w_c$ of the plant. This input sequence is a succession of commands sent by the controller to the actuators of the manufacturing system. As explained in Section 4.1.3, the controller is the first component of the control loop to send an output event based on the specification to achieve. In this case, the first target state is $\mathbf{Z_s}(2) = \{86\}$ with the degree of freedom $\mathbf{Z_s}(1) = *$ as defined in the specification (7.3). The computed control law in Fig. 7.8 requires that the controller sends the input sequence $W_c = (23, 22)$ to reach the state 86. Recall that the state 86 is the situation where there is a workpiece ready to be picked-up by the horizontal gripper, the magazine has 9 workpieces and the other components are in their respective initial state (see Tab. D.42). Thus, to reach the target state, the pusher must first push the workpieces on the buffer rail so that the very last workpiece is ready to be picked-up by the horizontal gripper. This is symbolized by the input event $\tilde{v}_p = 23$ in Tab. D.43. Next, the pusher is pulled back with the input event $\tilde{v}_p = 22$. In the mean time, the output sequence $\tilde{W}_p = (3, 86, 85, 87, 81)$ is registered from $t = 0$ to $t_{Z_s(2)} = 1.92 \ s$ in Fig. 7.9 by the sensors and is now interpreted with Tab. D.44 as follows:

- $\tilde{w}_p = 3$: all components are in their initial position, e.g., pusher retracted, no workpiece ready to be picked-up, magazine full, etc.

- $\tilde{w}_p = 86$: the pusher is blocking and the other components remain static. This event is recorded by a sensor placed between the fully retracted and fully extended position of the pusher. Hence, this event is recorded everything the pusher switches between the aforementioned positions.

- $\tilde{w}_p = 85$: the pusher is fully extended and there is one workpiece ready to be picked-up by the horizontal gripper. This is the plant output event expected by the controller as a response to $\tilde{v}_p = 23$.

- $\tilde{w}_p = 87$: the pusher is blocking and the other components remain static. As for $\tilde{w}_p = 49$, this event is recorded while the pusher is pulled back to the fully retracted position.

- $\tilde{w}_p = 81$: the pusher is fully retracted, the magazine has 9 workpieces and the other components are static. The number of workpieces in the magazine diminishes only now because the workpiece stack can only fall down on the buffer rail when the pusher is fully retracted. This event is expected by the controller as a response to $\tilde{v}_p = 22$.

The described interaction between the controller and the plant for the first two steps of the specification is the same during the whole process illustrated in Fig. 7.9. The state sequence $\hat{z}_p$ is derived from the internal states of $\mathcal{N}_c$ which are similar to those of the process $\mathcal{N}_p$ in the nominal case. This plot shows how all specified states are sequently reached until the final target state $Z_s(12) = 1$ at $t_{Z_s(12)} = 52.1\ s$. Therefore, the specification defined in (7.3) is fulfilled by the manufacturing system in the closed-loop with the controller $\mathcal{A}_{c-main}$ of Fig. 7.8.

In this section, only the fault free behavior of the manufacturing system was considered. In the following, two fault scenarios are assumed each in a specific subprocess of the main process. Section 7.4 handles the reconfiguration of the controller for the first subprocess under the presence of an actuator failure. Section 7.5 presents the reconfiguration of the controller for the second subprocess after a system internal fault.

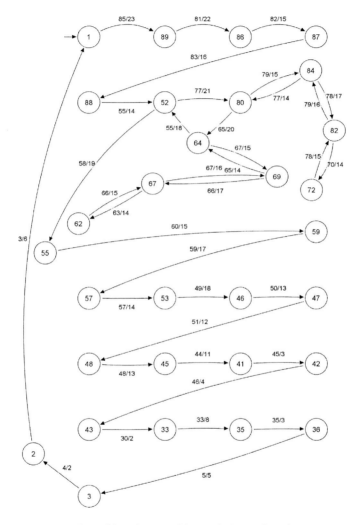

Figure 7.8.: Control law $\mathcal{A}_{c-main}$ of the nominal manufacturing process

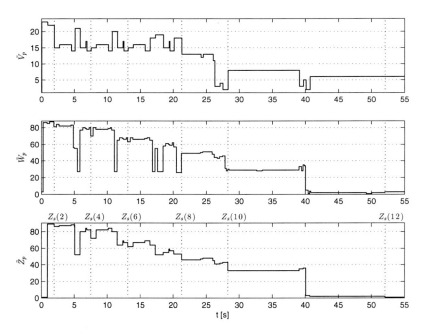

Figure 7.9.: Signal plot of the input, output and state sequences during the nominal manufacturing process

7.4. Control reconfiguration after an actuator failure

This section presents two fault scenarios which have been experimented on the manufacturing cell described in Section 7.1. The model of the complete process was built in Section 7.2. However, in order to simplify the investigations of the faulty behavior, the following sections focus solely on the respective subprocess with which is subject of faults. A subprocess here is seen as being part of the main process from Section 7.3. Hence, only the components concerned by the faults are modeled.

For the nominal and the faulty subprocess presented here, the model is explained the interaction with the controller are compared.

7.4.1. Nominal subprocess SP1

This part of the overall process concerns the transport of a workpiece from the dropping position of the conveyor belt to the state where the vertical gripper is holding that workpiece over the slide Sl1. In the main process specified in (7.1), only the steps from 8 to 10 are relevant. Thus, the specified vectorial sequence of the automata network shown in Fig. 7.6 for this subprocess is:

$$\mathcal{Z}_{s-SP1}(0\ldots2) =$$

	P	BR	M	HH	GHH	Pos7	Pos15	B	D1	D3	S1	S3	VH	GVH
0	1	1	2	1	1	1	1	2	1	1	1	1	1	1
1	*	*	*	*	*	*	*	*	*	*	*	*	*	*
2	1	1	2	1	1	1	1	1	1	1	1	1	1	2

$$\tag{7.4}$$

In (7.4), note that the conveyor belt in the B-column of $\mathcal{Z}_{s-SP1}(0)$ is activated at state 2. The state 2 in the GVH-column of $\mathcal{Z}_{s-SP1}(2)$ means that the vertical gripper is over the deflector D1 holding a workpiece. The meaning of the symbols are given in Section D.7 and Section D.11. The encoded state specification sequence is

$$\mathcal{Z}_s(0\ldots2) = (46, *, 33)^T. \tag{7.5}$$

Note also that the degree of freedom in $\mathcal{Z}_{s-SP1}(1)$ do not specify which slide to use. This will be fixed by the computed control law $\mathcal{A}_{c-SP1}$ resulting from the model of the manufacturing system and the specification defined in (7.4).

The resulting control law of this subprocess is depicted in Fig. 7.10.

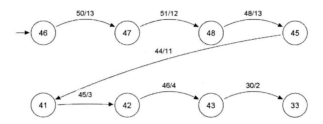

Figure 7.10.: Control law $\mathcal{A}_{c-SP1}$ of the subprocess 1 in the manufacturing cell

Section D.12 contains the encoding tables of the all states and signals used in $\mathcal{A}_{c-SP1}$. The internal state sequence of the controller $\mathcal{A}_{c-SP1}$ is

$$\tilde{Z}_{c-SP1} = (46, 47, 48, 45, 41, 42, 43, 33). \tag{7.6}$$

The control set sequence $\tilde{Z}_{c-SP1}$ is an element of the set of state sequences represented by the specification defined in (7.5) where the state sequence $(47, 48, 45, 41, 42, 43)$ in (7.6) represents a selected trajectory from the degree of freedom specified as $*$ in (7.5). The decoded internal state sequence of the controller $\mathcal{A}_{c-SP1}$ is

$Z_{c-SP1}(0 \ldots 7) =$

	P	BR	M	HH	GHH	Pos7	Pos15	B	D1	D3	S1	S3	VH	GVH
0	1	1	2	1	1	1	1	2	1	1	1	1	1	1
1	1	1	2	1	1	1	1	3	1	1	1	1	1	1
2	1	1	2	1	1	1	1	3	2	1	1	1	1	1
3	1	1	2	1	1	1	1	1	2	1	2	1	1	1
4	1	1	2	1	1	1	1	1	1	1	2	1	1	1
5	1	1	2	1	1	1	1	1	1	1	2	1	1	3
6	1	1	2	1	1	1	1	1	1	1	2	1	1	4
7	1	1	2	1	1	1	1	1	1	1	1	1	1	2

$$\tag{7.7}$$

The state sequences of relevant and irrelevant components for this subprocess can be distinguished in (7.7). Irrelevant components have their columns filled with the same symbol (1 or 2) except the VH-column which though represent a relevant component in this subprocess. Other relevant component are the belt (B), deflector 1 (D1), slide 1 (S1) and the vertical gripper (GVH). The first three lines of the B-column with the transition $2 \rightarrow 3$ show the sequential activation of sensor 1 and sensor 2 along the belt (see Section D.7). Next, the transition $1 \rightarrow 2$ in deflector 1 and slide 1 respectively reveal the oblique position of the deflector and the presence of a workpiece in slide 1 (see Section D.8 and D.9). The last three lines of the GVH-column show the steps in which the workpiece is picked-up from slide 1 and held above (see Section D.11). The fact that the VH-column only contains ones is due to the fact that the state 1 of the vertical handler stands for the position where the whole subprocess is taking place.

The interaction of the plant with the controller for this subprocess is depicted in the plots of Fig. 7.11. The input and output sequences recorded in the first and second plot respectively are reflected along the I/O transitions of the control law $\mathcal{A}_{c-SP1}$ in Fig. 7.10 from state 46 to state 33. The third plot represents the state sequence given in (7.6). Hence, the specification defined in (7.5) is obviously fulfilled by the manufacturing cell with the controller $\mathcal{A}_{c-SP1}$.

7.4.2. Actuator failure

Now the situation where the controller $\mathcal{A}_{c-SP1}$ (Fig. 7.10) of the subprocess SP1 tries to move deflector 1 into its oblique position but the deflector remains straight. This happens when the controller is at state $z = 47$ decoded w.r.t Tab. D.42 on page 285 into

$$z_{c-SP1} =$$

P	BR	M	HH	GHH	Pos7	Pos15	B	D1	D3	S1	S3	VH	GVH
$\big(1$	1	2	1	1	1	1	3	1	1	1	1	1	$1 \big)$

$$(7.8)$$

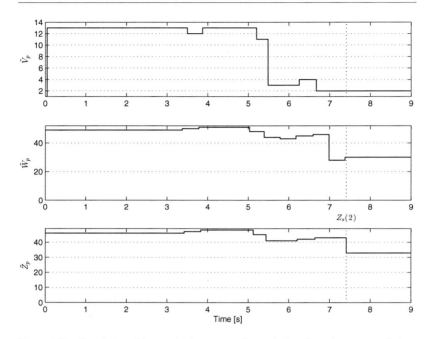

Figure 7.11.: Signal plot of the nominal transport of a workpiece from the conveyor belt to slide 1

and sends the output 12 decoded w.r.t. Tab. D.44 into

$$\boldsymbol{w}_{c-SP1} =$$

P	BR	M	HH	GHH	Pos7	Pos15	B	D1	D3	S1	S3	VH	GVH
0	0	0	0	0	0	0	0	2	0	0	0	0	0

$$(7.9)$$

The B-column of (7.8) shows that there is a workpiece detected next to slide 1 and the D1-column of (7.10) shows that the controller activates the oblique position of deflector 1.

Based on the control law depicted in Fig. 7.10, the controller expects the input 51 from the plant decoded w.r.t. Tab. D.43 as

$$v_{c-SP1} =$$

P	BR	M	HH	GHH	Pos7	Pos15	B	D1	D3	S1	S3	VH	GVH
1	1	3	1	1	0	0	3	2	1	2	2	1	1

$$(7.10)$$

showing that the deflector 1 is oblique and the slides are still empty (see Sections D.8 and D.9).

However, the controller sequentially receives the input 50 as shown in the second plot of Fig. 7.14. $\tilde{w}_p = 50$ means that deflector 1 is still in its straight position. The input and state sequences $\tilde{V}_p$ and $\tilde{Z}_p$ both show that the control loop is blocking and the specification defined in (7.5) is not achieved from time $t_f = 4.9\ s$ to $t_r = 15.2\ s$ and even beyond until $t_n = 36\ s$. Without control reconfiguration, the workpiece does not slip into slide 1 but is transported out the manufacturing cell by the conveyor belt. The reconfiguration of the controller for this actuator failure is presented now.

7.4.3. Fault-tolerant control with a deflector failure

Reconfigurability. First, the reconfigurability of the controller for the considered failure has to be checked before attempting a control reconfiguration. According to Theorem 5.1 on page 120, there are two conditions to satisfy in order to guarantee reconfigurability:

1. Consider the subgraph supercontroller $\mathcal{N}_{c-SP1}$ of subprocess 1 depicted in Fig. 7.12. The redundancy degree of $\mathcal{N}_{c-SP1}$ is obtained by applying (5.18) which leads to $RD(\mathcal{N}_{c-SP1}) \geq 1 > 0$ and satisfies the first condition of Theorem 5.1.

2. The second condition requires the safe feasibility of the specification in the faulty plant. Recall that the current target state is 33 according to the specification as defined in (7.5). The model of the faulty plant is not depicted here due to the state space complexity. But since the specified target state is reachable from the initial state 12 in the supercontroller, it is also reachable in the nominal plant since the supercontroller stems from it. This is still true for the faulty plant because the deflector 1 failure only make one state sequence to state 1 invalid but not all existing state sequences to state

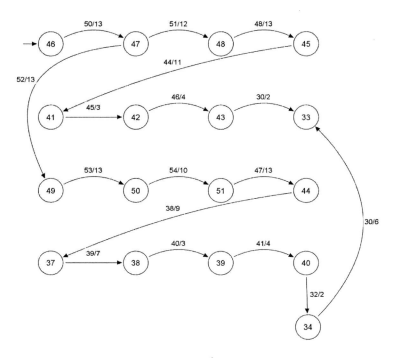

Figure 7.12.: Subgraph of the supercontroller $\mathcal{N}_{c-SP1}$ of the subprocess 1 in the manufacturing cell

1. Hence, the specification is still feasible. In addition, the fact that the supercontroller is output-deterministic ensures the safety condition because no deviation from the specified state sequence can be inadvertently provoked by the controller but the faulty plant only. Therefore, the specification is safely feasible in the faulty plant and the second condition of Theorem 5.1 is fulfilled.

Forward control reconfiguration of $\mathcal{A}_{c-SP1}$**.** Now that the reconfigurability of the controller is formally given, the control reconfiguration is executed w.r.t Algorithm 3 as follows. Among other information the on-line control reconfiguration algorithm has the state sequence $\bar{Z}_s(0 \cdots k_e) = (46, 47, 48, 45, 41, 42, 43, 33)$. At the moment of fault occurrence denoted k_{sf}, the controller is blocking at state $\bar{Z}_s(k_{sf}) = 47$ and the process is temporarily stopped by the reconfiguration unit with an ε symbol suggested in (5.6). In Fig. 7.14, the ε symbol is represented by $\tilde{v}_p = 1$ which means that every component of the network in Fig. 7.6 receives an ε coded as $v_\iota = 0$, $\iota = 1, \ldots, \mu$, in Tab. D.43. Since the controller is blocking at the same time t_f depicted in Fig. 7.14, the forward reconfiguration is launched. The next specified state is $\bar{Z}_s(k_{sf} + 1) = 48$. Thus, a Breadth-First-Search is performed in $\mathcal{N}_{c-SP1}$ and formalized by $\bar{Z}_{\text{BFS}} = \text{BFS}(47, 48, \mathcal{N}_{c-SP1})$. The index of the target state in the BFS($\cdot$) operator is incremented in $\bar{Z}_s(0 \cdots k_e)$ until a path is found and $\bar{Z}_{\text{BFS}}$ is nonempty. That is the case for $\bar{Z}_{\text{BFS}} = \text{BFS}(47, 33, \mathcal{N}_{c-SP1})$ where $\bar{Z}_{\text{BFS}} = (47, 49, 50, 51, 44, 37, 38, 39, 40, 34, 33)$. Next, the corresponding inputs and outputs of the new state sequence are read from the supercontroller $\mathcal{N}_{c-SP1}$ in Fig. 7.12. Each new transition is inserted in the set L_{c-set1} and included in the reconfigured control law $\mathcal{A}^r_{c-SP1}$ depicted in Fig. 7.13.

The signal plots of the reconfigured control loop are depicted in Fig. 7.14. The solid lines represent the reconfigured behavior and the dashed lines represent the nominal behavior synchronized with the former to simply the comparison of both. The three vertical lines t_f, t_r and t_n show the occurrence of the following events:

- t_f: the time where the fault is diagnosed and the reconfiguration starts.

- t_r: the time where the controller reconfiguration is finished and the new control law is applied.

- t_n: the time where the behavior of the reconfigured control loop matches with the nominal control loop. At this time, the specification is fulfilled.

All three plots of the input, output and state sequence of Fig. 7.14 show that the dynamic of the control loop is stopped from $t_f = 4.9\ s$ to $t_r = 15.2\ s$. In this time, the conveyor belt is stopped with $v_p = 1$ which represents the ε symbol. At $t_r = 15.2\ s$ the reconfiguration is finished and the new control law depicted in Fig. 7.13 is applied. The new transition goes from state 47 to state 49 with the I/O label 52/13. Thus, the input signal $\tilde{v}_p = 13$ which restarts the conveyor belt is set at t_r in the first plot of Fig. 7.14. This input signal is maintained until the plant reaches state $\tilde{z} = 50$ meaning that the workpiece is detected by sensor 4, i.e, next to slide 3. Recall that sensor 1 is at the dropping position of the conveyor

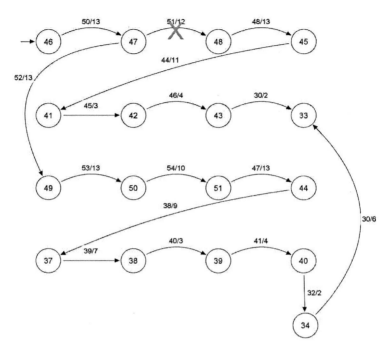

Figure 7.13.: Reconfigured controller $\mathcal{A}^r_{c-SP1}$ of the subprocess 1 in the manufacturing cell after a deflector failure

belt. Now, the deflector 3 is activated with $\tilde{v}_p = 10$ and the plant reacts with the value $\tilde{w}_p = 54$ decoded into

$$v^r_{c-SP1} =$$

P	BR	M	HH	GHH	Pos7	Pos15	B	D1	D3	S1	S3	VH	GVH
1	1	3	1	1	0	0	5	1	2	2	2	1	1

$$(7.11)$$

Equation (7.11) shows that deflector 3 is in its oblique position, so that the workpiece will slip into slide 3 as reflected by the next output $\tilde{w}_p = 47$ (see Tab. D.44, Section D.8 and Section D.9). Thus, the output sequence $\tilde{W}_p = (54, 47)$ shows how deflector 1 is replaced

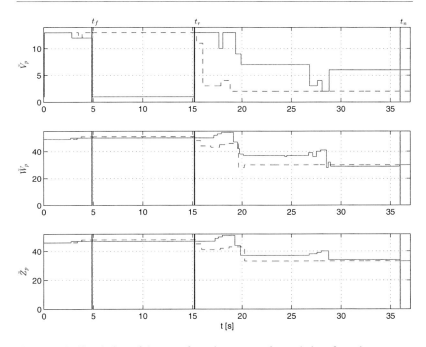

Figure 7.14.: Signal plots of the reconfigured transport of a workpiece from the conveyor belt to slide 1 with deflector failure

by deflector 3 after the controller reconfiguration. The subsequent steps show that the vertical handler moves to position 3, the vertical gripper picks up the workpiece from slide 3 and transport it to position 1 above slide 1. At time $t_n = 36\ s$ (Fig. 7.14) the reconfigured behavior and the nominal behavior match because the plant reaches the target state 1 which was defined by the specification (7.5). Therefore, the faulty manufacturing cell still fulfills the specification despite the broken deflector 1 with the reconfigured controller $\mathcal{A}^r_{c-SP1}$.

7.5. Control reconfiguration after a system internal fault

7.5.1. Nominal subprocess SP2

This subprocess is the third step of the main process specified in Section 7.3. It concerns the case where the horizontal gripper GHH has to transport a workpiece from position 15 to position 7. In the main process specified in (7.1), only the steps from 4 to 6 are relevant. Thus, the specified vectorial sequence of the automata network shown in Fig. 7.6 for this subprocess is:

$$\mathcal{Z}_{s-SP2}(0\ldots 2) =$$

	P	BR	M	HH	GHH	Pos7	Pos15	B	D1	D3	S1	S3	VH	GVH
0	1	1	2	15	1	1	2	1	1	1	1	1	1	1
1	*	*	*	*	*	*	*	*	*	*	*	*	*	*
2	1	1	2	7	1	2	1	1	1	1	1	1	1	1

$$\tag{7.12}$$

Note that the verbal specification stated above for this subprocess is reflected by the columns of the only components concerned so far in (7.12): HH, GHH, Pos7 and Pos15. The HH-column explicitly shows the transition of the horizontal handler from position 15 to position 7. The GHH-column shows that the horizontal handling gripper starts and ends at state $z_{GHH} = 1$ with intermediary states symbolized by the $*$ symbol. $z_{GHH} = 1$ means that the horizontal gripper is up without workpiece (see Tab. D.17). Further more, observe that z_{Pos15} moves from state 2 which means there is a workpiece inside, to state 1 which means there is no workpiece inside (see Tab. D.21). The fact that z_{Pos7} performs the opposite transition, i.e, from state 1 to state 2 means that the workpiece moved from position 15 to position 7.

Based on the state encoding table given in Section D.12, the encoded specified state sequence of (7.12) is

$$\mathcal{Z}_s(0\ldots 2) = (72, *, 62)^T. \tag{7.13}$$

Due to the $*$ symbol in (7.13), the exact trajectory to be followed from state 72 to state 62 is not specified. This is fixed by the resulting control law $\mathcal{A}_{c-SP2}$ depicted in Fig. 7.15. The internal state sequence of the controller $\mathcal{A}_{c-SP2}$ is

$$\mathbf{Z}_{c-SP2} = (72, 82, 84, 80, 64, 69, 67, 62). \tag{7.14}$$

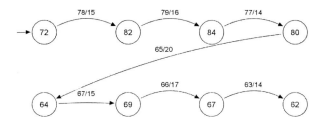

Figure 7.15.: Control law $\mathcal{A}_{c-SP2}$ of the subprocess 2 in the manufacturing cell

The encoding tables of the all states and signals used in $\mathcal{A}_{c-SP2}$ are given in Section D.12. The control set sequence Z_{c-SP2} is an element of the set of state sequences represented by the specification defined in (7.13) where the state sequence $(82, 84, 80, 64, 69, 67)$ in (7.14) represents a selected trajectory from the degree of freedom specified in (7.13). The decoded internal state sequence of the controller $\mathcal{A}_{c-SP2}$ is

$$Z_{c-SP2}(0\ldots 7) =$$

	P	BR	M	HH	GHH	Pos7	Pos15	B	D1	D3	S1	S3	VH	GVH
0	1	1	2	15	1	1	2	1	1	1	1	1	1	1
1	1	1	2	15	3	1	2	1	1	1	1	1	1	1
2	1	1	2	15	4	1	2	1	1	1	1	1	1	1
3	1	1	2	15	2	1	1	1	1	1	1	1	1	1
4	1	1	2	7	2	1	1	1	1	1	1	1	1	1
5	1	1	2	7	4	2	1	1	1	1	1	1	1	1
6	1	1	2	7	3	2	1	1	1	1	1	1	1	1
7	1	1	2	7	1	2	1	1	1	1	1	1	1	1

$$(7.15)$$

As in the first subprocess, irrelevant components can be distinguished here through their columns filled with ones since they remain in their respective initial state. The four components horizontal handler (HH), horizontal gripper (GHH), position 7 (Pos7) and position 15 (Pos15) explicitly reflect the required dynamics for the second subprocess. From step $k = 0$ to $k = 3$ in (7.15) executed by the controller, the state sequences Z_{GHH} and Z_{Pos15} show that the horizontal gripper picks up the workpiece as described in Tab. D.17 and Tab.

D.21 respectively. At step $k = 4$, the horizontal gripper is holding the workpiece over position 7 which is empty as position 15. From step $k = 5$ to $k = 7$, the workpiece is dropped by the horizontal gripper in position 7 and the specification of the second subprocess is fulfilled.

The interaction of the plant with the controller for this subprocess is depicted in the plots of Fig. 7.16. As in the first subprocess, the input, output and state sequences recorded in

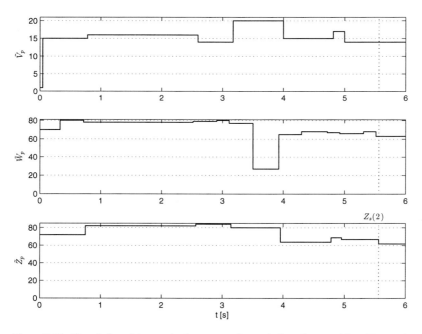

Figure 7.16.: Signal plot of the nominal transport of a workpiece from position 15 to position 7

the first and second plot are respectively reflected along the I/O transitions of the control law $\mathcal{A}_{c-SP2}$ in Fig. 7.15 from state 72 to state 62. The third plot represents the state sequence given in (7.14). Hence, the specification defined in (7.13) is obviously fulfilled by the manufacturing cell with the controller $\mathcal{A}_{c-SP2}$.

7.5.2. System internal fault

The system internal fault is provoked by placing a workpiece in position 7 before the sub-process is started. Thus, when the subprocess is started, the nominal controller expects position 7 to be empty whereas it is already occupied. As a consequence, the gripper will not be able to drop the workpiece correctly as in the nominal case. This situation occurs when the controller is at state $z_{c-SP2} = 64$ decoded with Tab. D.42 into

$z_{c-SP2} =$

P	BR	M	HH	GHH	Pos7	Pos15	B	D1	D3	S1	S3	VH	GVH
$\left(1\right.$	1	2	7	2	1	1	1	1	1	1	1	1	$1 \left.\right)$

$$(7.16)$$

and sends the output 15 decoded w.r.t. Tab. D.44 into

$w_{c-SP2} =$

P	BR	M	HH	GHH	Pos7	Pos15	B	D1	D3	S1	S3	VH	GVH
$\left(0\right.$	0	0	0	2	0	0	0	0	0	0	0	0	$0 \left.\right).$

$$(7.17)$$

Equation (7.17) shows that the horizontal gripper receives the command

$$v_{c-GHH} = w_{c-GHH} = 2$$

which is a command to go down w.r.t. Tab. D.17 in order to drop the workpiece. The controller expects the value $\hat{v}_{c-SP2} = 67$ which confirms that the horizontal gripper is down with the workpiece ready to be dropped by turning the vacuum off in the next step with $w_{c-SP2} = 17$ i.e $w_{c-GHH} = 4$ (see Section D.12 to decode these values). Instead, the plant reacts with the output $\tilde{w}_p^f = 68$ decoded as

$w_{p-SP2}^f =$

P	BR	M	HH	GHH	Pos7	Pos15	B	D1	D3	S1	S3	VH	GVH
$\left(1\right.$	1	3	7	5	0	0	1	1	1	2	2	1	$1 \left.\right)$

$$(7.18)$$

showing through $w_{p-GHH} = 5$ that the workpiece can not be dropped. The fact a different output than the one expected, was generated by the plant shows that it has moved to a faulty state. The diagnosis unit reports that the plant is now at state $\tilde{z}^f_{p-SP2} = 71$ decoded into

$$
z^f_{p-SP2} =
$$

P	BR	M	HH	GHH	Pos7	Pos15	B	D1	D3	S1	S3	VH	GVH
$\big(1$	1	2	7	5	2	1	1	1	1	1	1	1	$1\big)$

$$(7.19)$$

where the nominal controller $\mathcal{A}_{c-SP2}$ is also blocking. Note that $z_{p-Pos7} = 2$ is obtained by the diagnosis unit and means that there is a workpiece in position 7 according to Tab. D.21. This blocking situation is illustrated in Fig. 7.17 from $t_f = 6.4\ s$ to $t_r = 93.2\ s$ where the nominal and the faulty behavior with subsequent reconfiguration have been synchronized for analysis purposes. The dashed lines represent the nominal behavior and the solid lines represent the faulty behavior with subsequent reconfiguration. Recall that t_f is the time where the fault is diagnosed and the reconfiguration is launched. At time t_r the reconfigured controller is ready and takes correcting actions.

During the time interval $[t_f, t_r]$, the plant permanently receives the input $\tilde{v}_p = 1$ which represents the ε symbol for which the command of all actuators is set to 0 (see Tab. D.43). In this way, the own dynamic of the plant is stopped as suggested in (5.6). The second plot (in dashed lines) shows the almost overlapping expected nominal output $w^n_p = 67$ and the faulty output $w^f_p = 68$ (in solid lines). Finally, the third plot clearly shows a discrepancy from t_f to t_r between the state $\tilde{z}^n_p = 62$ where the plant should be by the time t_f and the faulty state state $z^f_p = 71$ where is has moved into.

The next paragraph explains how the controller is reconfigured for this system internal fault.

7.5.3. Fault-tolerant control despite an occupied position

Reconfigurability. First, the two reconfigurability conditions of Theorem 5.1 are now investigated:

1. Consider the subgraph of the supercontroller $\mathcal{N}_{c-SP2}$ of subprocess 2 depicted in Fig. 7.18. The redundancy degree of $\mathcal{N}_{c-SP2}$ is obtained by applying (5.18) which leads to $rd(\mathcal{N}_{c-SP2}) \geq 1 > 0$ and satisfies the first condition of Theorem 5.1.

2. The safe feasibility of the specification in the faulty plant needs to be investigated

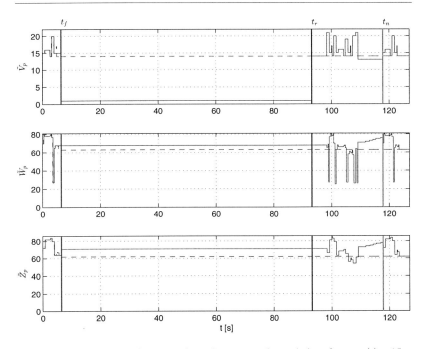

Figure 7.17.: Signal plots of the reconfigured transport of a workpiece from position 15 to position 7

from state $z^f_{p-SP2} = 64$ in Fig. 7.18. This state is critical for the feasibility of the specification due to the nondeterministic transitions of the faulty plant $\mathcal{N}^f_{p-SP2}$, namely, $(69, 67, 64, 15)$ and $(71, 68, 64, 15)$. Note that the latter transition is in dashed lines in Fig. 7.18 because it can be achieved by the plant alone but is provoked by an external mean, i.e, adding a workpiece into the position 7 where the horizontal handler is moving to. This means that the nondeterminism of the faulty plant is provoked from outside the plant. Thus, the transition $(71, 68, 64, 15)$ has an obsolete character and can not be executed unless another workpiece is added from outside. Since this critical transition does not exists in the "intrinsic" behavior of the plant, it is excluded from the feasibility analysis. The specification is then obviously safely feasible in the faulty plant and the second condition of Theorem 5.1 is fulfilled.

Backward control reconfiguration of $\mathcal{A}_{c-SP2}$. The control reconfiguration is now executed w.r.t Algorithm 3 as follows. The state sequence stored in the nominal controller $\mathcal{A}_{c-SP2}$ is $\bar{Z}_s(0 \cdots k_e) = (72, 82, 84, 80, 64, 69, 67, 62)$. At the moment of diagnosis denoted k_{sf} in Algorithm 3 and t_f in Fig. 7.17, the nominal controller is blocking at state $\bar{Z}_s(k_{sf}) = 64$ but the plant has moved to the state $Z_p(k_f) = 71$. However, the moment of the fault occurrence lies in the past since the additional workpiece was added even before the subprocess was started. Hence, $k_{sf} < k_s$ holds and the backward reconfiguration is launched. The first attempt of setting the characteristic function to one for the transition $71 \rightarrow 64$ fails because there is no such a transition in the faulty plant. The algorithm proceeds with the computation of the new state trajectory by means of the Breadth-First-Search in $\mathcal{N}_{c-SP2}$ from state 71 subsequently to the states 80, 84, 82 and 72. Only the latter leads to a nonempty state sequence formalized as $\bar{Z}_{\text{BFS}} = \text{BFS}(71, 72, \mathcal{N}_{c-SP2}) = (66, 81, \ldots, 78)$ depicted in Fig. 7.18. The corresponding inputs and outputs of the new state sequence are read from the supercontroller $\mathcal{N}_{c-SP2}$ in Fig. 7.18. Each new transition is inserted in the set L_{c-set1} and included in the reconfigured control law $\mathcal{A}_{c-SP2}^r$ depicted in Fig. 7.19. $\mathcal{A}_{c-SP2}^r$ differs from $\mathcal{N}_{c-SP2}$ by the initial state which is now $z_{0c-SP2}^r = 71$ and the state transition $64 \rightarrow 71$ which was performed by the faulty plant because of the workpiece added before the process was started.

Recall that Fig. 7.18 solely shows a subgraph of the supercontroller. The complete supercontroller graph has far more other transitions where the plant could switch to and still be able to reach the target state 62 with a reconfigured controller. On the other hand, should the plant reach a failure state z_ε which was not modeled, then the reconfigurability condition would be violated because the specification would not be achievable since the specification automaton would be empty in that case.

The signal plots of the reconfigured control loop are depicted in Fig. 7.17. The meaning of the vertical time lines marked as t_f, t_r and t_n is given on page 159. The behavior of the control loop from t_f to t_r has already been explained in the previous section. Now, the behavior of the control loop is explained from t_r to t_n only. This time period is particularly zoomed in Fig. 7.20. The newly computed state sequence during the reconfiguration can be recognized in the solid lines of the third plot of Fig. 7.20. After the ε command, the first command sent out by the reconfigured controller to the plant is $\tilde{v}_{p-SP2} = 14$ which means to pull the horizontal gripper back to the state "up with a workpiece over position 7" which is the state $z_{p-SP2} = 66$. In order to be able to interpret these numbers as well as those used in the sequel, it is required to make use of the tables given in Appendix D. Only the critical states and events will be explained next. The set of state sequences are now labeled on the left hand side with the corresponding encoded state instead of the corresponding steps as presented so far. The main correcting actions undertaken by the reconfigured controller are the following:

1. $[t_r, t_{r1}]$: Move the workpiece held by the horizontal gripper back to position 15. The corresponding state sequence goes from state 71 to state 79 and is decoded as

$Z^f_{p-SP2} =$

	P	BR	M	HH	GHH	Pos7	Pos15	B	D1	D3	S1	S3	VH	GVH
71	1	1	2	7	5	2	1	1	1	1	1	1	1	1
66	1	1	2	7	2	2	1	1	1	1	1	1	1	1
81	1	1	2	15	2	2	1	1	1	1	1	1	1	1
85	1	1	2	15	4	2	2	1	1	1	1	1	1	1
83	1	1	2	15	3	2	2	1	1	1	1	1	1	1
79	1	1	2	15	1	2	2	1	1	1	1	1	1	1

$$(7.20)$$

Observe the columns of the horizontal handler (HH) and its gripper (GHH). They reflect the transport of the workpiece to position 15, the subsequent movements of the gripper including the activation and deactivation of vacuum.

2. $[t_{r1}, t_{r2}]$: Go to position 7 and pick up the wrong workpiece. The corresponding state sequence goes from state 79 to state 65 and is decoded as

$Z^f_{p-SP2} =$

	P	BR	M	HH	GHH	Pos7	Pos15	B	D1	D3	S1	S3	VH	GVH
79	1	1	2	15	1	2	2	1	1	1	1	1	1	1
63	1	1	2	7	1	2	2	1	1	1	1	1	1	1
68	1	1	2	7	3	2	2	1	1	1	1	1	1	1
70	1	1	2	7	4	2	2	1	1	1	1	1	1	1
65	1	1	2	7	2	1	2	1	1	1	1	1	1	1

$$(7.21)$$

Observe the columns of the horizontal handler (HH) and its gripper (GHH). They reflect the move of the horizontal handler to position 7, the subsequent movements of the gripper including the activation of vacuum. At state 65, the gripper is holding the workpiece which is responsible for the blocking situation over position 7.

3. $[t_{r2}, t_{r3}]$: Go to the conveyor belt and drop the wrong workpiece. The corresponding state sequence goes from state 65 to state 54 and is decoded as

$Z^f_{p-SP2} =$

	P	BR	M	HH	GHH	Pos7	Pos15	B	D1	D3	S1	S3	VH	GVH
65	1	1	2	7	2	1	2	1	1	1	1	1	1	1
56	1	1	2	2	2	1	2	1	1	1	1	1	1	1
60	1	1	2	2	4	1	2	2	1	1	1	1	1	1
58	1	1	2	2	3	1	2	2	1	1	1	1	1	1
54	1	1	2	2	1	1	2	2	1	1	1	1	1	1

$$(7.22)$$

This sequence is similar to the first one whereas the only difference resides in the position where the workpiece is being dropped, i.e, the conveyor belt (B), namely in position 2 in this step.

4. $[t_{r3}, t_{r4}]$: Go back to position 15 and wait until the wrong workpiece is out of the system. The corresponding state sequence goes from state 54 to state 78 and is decoded as

$Z^f_{p-SP2} =$

	P	BR	M	HH	GHH	Pos7	Pos15	B	D1	D3	S1	S3	VH	GVH
54	1	1	2	2	1	1	2	2	1	1	1	1	1	1
61	1	1	2	7	1	1	2	2	1	1	1	1	1	1
73	1	1	2	15	1	1	2	2	1	1	1	1	1	1
74	1	1	2	15	1	1	2	3	1	1	1	1	1	1
75	1	1	2	15	1	1	2	4	1	1	1	1	1	1
76	1	1	2	15	1	1	2	5	1	1	1	1	1	1
77	1	1	2	15	1	1	2	6	1	1	1	1	1	1
78	1	1	2	15	1	1	2	7	1	1	1	1	1	1

$$(7.23)$$

Observe that the horizontal handler (HH) first move to position 7 before moving to position 15. The intermediate stop on state 7 is only due to the fact that the currently executed state sequence is the first one found by the reconfiguration algorithm. There is no technical advantage for this. However, the conveyor belt (B) is activated from state 73 on with the input $\tilde{v}^f_{p-SP2} = 13$ and its state evolve from 2 to 7. These shows that the workpiece is subsequently being detected by the sensors along the conveyor belt. In Fig. 7.20, this period is characterized by the stairs in the output and the state sequence plot.

5. $[t_{r4}, t_n]$: Continue with the process as in the nominal case. The remaining states, output and input transitions coincide with those of the nominal behavior of Fig. 7.16. The decoded state sequences in this last part of the reconfigured behavior are identical with (7.15).

7.6. Experimental limitations and sketches of solutions

Limitations of the control reconfiguration are encountered due to the following reasons:

- **Complexity of the model.** This is a well known problem in the discrete-event systems community. It is the reason why the search of new trajectory often represents the bottle neck of reconfiguration algorithms. There exists theoretical search algorithms which guarantee to find a path from an initial to a target node of graph as long as it exists. Some of them even guarantee to find the optimal path in terms of cost and length. However, for practical use, some heuristics are necessary to guide the search algorithm in the right direction otherwise, e.g., hard real-time constraints can not be met. This results in a so called informed search [174]. However, with growing complexity, even the best search algorithm can not always help. The search problem itself needs to be simplified as far as possible.

- **Nonideal Communication.** The bandwidth and the data rate are assumed to be infinitely large in theoretical approaches. In a practical environment, the communication channel also plays a key role in the reconfiguration. However, the requirements to be put on the communication channel depend on the maximal amount of data to be exchanged and the smallest transition time in a graph. In the considered examples the amount of data was low. However, the fluid level control system has a slower dynamic, thus, does not required a channel with high data rate. This is not the case for the manufacturing process where the shortest and most time critical transition is the time left to transport a workpiece from the belt to another slides once the expected slides was missed.

- **Limited motion of the physical parts.** This is due to either because of their own shape or because some components might be obstacle to others.

Possible sketches of solutions to reduce the complexity of the search algorithms or the search space:

- Use of symmetries as far as there are known in advance.

- While modeling, consider grouping elements, equivalent states or simply run a minimization algorithm.

- Exploit the limited reconfiguration possibilities of the supercontroller $\mathcal{N}_c$. A similar approach was proposed in [183, 184] for self-reconfigurable robots. This approach sounds promising on one hand. On the other hand it shows limitations in cases where the faulty plant reaches states which were forbidden by the specification, hence, not included in the controller. Thus, in the case they are reached by the plant a reconfiguration sequence enabling to drive the plant back to a safe state does not exist in $\mathcal{N}_c^f$. For these cases, it might be suitable to augment all the states of $\mathcal{N}_p$ from which any state of $\mathcal{N}_c$ can be reached and obtain a robust controller $\mathcal{N}_{rc}$ as experienced in [25].

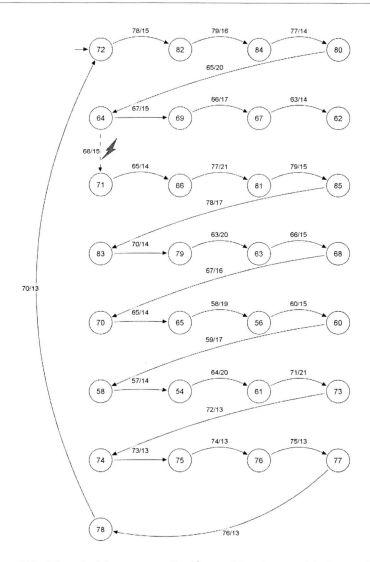

Figure 7.18.: Subgraph of the supercontroller $\mathcal{N}_{c-SP2}$ of the subprocess 2 in the manufacturing cell

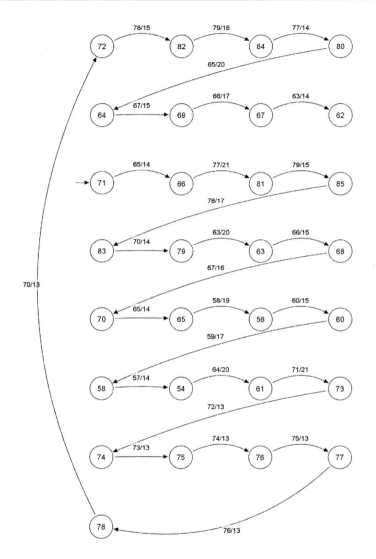

Figure 7.19.: Reconfigured controller $\mathcal{A}^r_{c-SP2}$ of the subprocess 2 in the manufacturing cell with a wrongly occupied position

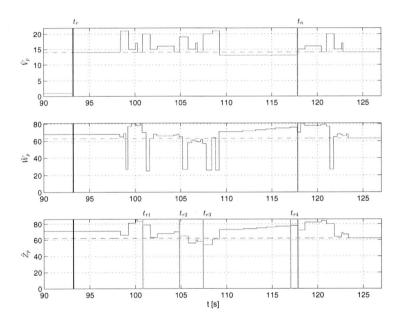

Figure 7.20.: Zoomed signal plots of the reconfigured transport of a workpiece from position 15 to position 7

8. Conclusion

8.1. Summary

This thesis has presented novel control reconfiguration methods and reconfigurability conditions for actuator failures, actuator faults and sensor faults. The reconfiguration methods include trajectory re-planning and fault-hiding approaches in the discrete-event framework of nondeterministic I/O automata. A new formalism to analyze the control loop, a reconfigurable controller architecture and error maps describing the fault have been introduced. Intermediary results which are related to each component of Fig. 1.1, i.e., the plant, the controller, the faults, and the control reconfiguration of have been obtained.

The formalism based on the analysis tools developed in Section 2.2 has been used throughout the resolution of the control reconfiguration problem for nondeterministic I/O automata:

- Active inputs, outputs and next states operators have been shown to be suitable to select particular events or states in an I/O automaton. Their usability was experienced in theorems, algorithms or proofs in combination with characteristic functions and predicate logic expressions.

- The newly introduced transition degree has been used to determine the number of possible control laws for a specification. This property has been exploited in terms of adjacency matrices to define a redundancy degree which is crucial for reconfigurability under actuator failures.

- The visualization of the dynamic behavior of I/O automata is adequately reflected by the I/O trellis automata. The formalization of the unfolding procedure through which an I/O trellis automaton is derived from a classic I/O automaton has been given in this thesis.

These formal extensions served to describe the nominal and faulty plant's behavior alone and in a closed-loop with a controller.

The control design method presented in Chapter 4 offers the possibility to build an autonomous control loop for a given specification. The safety condition has been expressed in terms of safe feasibility and included in the necessary part of reconfigurability conditions. For a safely feasible specification, the controllability condition was defined w.r.t. the W-determinism of the controller as a key property. This property insures the weak well-posedness of the control loop where an output deterministic controller is still able to guarantee the fulfillment of a specification in a control loop with a nondeterministic plant. The internal structure of a controller developed in this thesis captures the behavior of blocking and nonblocking control loop through the comparative counter. After a fault, this counter is not incremented and the reconfiguration begins.

The reconfiguration Algorithm 3 requires a model of the faulty plant. The latter is systematically obtained in this thesis through error maps at the input, output and system internal level. The signal falsification paradigm developed here has been used to determine the characteristic function of a faulty behavior based on the nominal behavior and the considered error maps. The application of this method on graphs of I/O automata results in self-loops, wrong state transition including possible overlapping with nominal state transitions. These overlapping transitions are critical for the reconfigurability and diagnosability because of emerging ambiguities. The reconfigurability criteria proposed here make use of the safe feasibility condition to solve the aforementioned ambiguous cases. For actuator failures, the reconfigurability condition confirms the heuristic requirement of a physical redundancy formalized in this thesis by the nonvanishing redundancy degree of the super-controller.

The main reconfiguration techniques formalized in this thesis are the trajectory re-planning and the I/O adaptation. Trajectory re-planning is formally achieved through symbolic states adjacency matrices. Due to the complexity of certain plants such as the pilot manufacturing cell, a Breadth-First-Search has been applied instead of adjacency matrices, especially for on-line reconfiguration where resources are limited. The I/O adaptation has been achieved in an implicit and an explicit way:

- The implicit I/O adaptation occurs after a trajectory re-planning when the corresponding input and output events are inserted in the list of transitions to be set or deleted in order to build the reconfigured controller.

- The explicit I/O adaptation can only be achieved if the corresponding error map of the concerned event is available. This is the situation where physical redundancies are not required since the reconfiguration is merely performed analytically.

These reconfiguration techniques have been demonstrated for off-line and on-line reconfiguration. Two on-line reconfiguration paradigms have been studied here: the forward and backward control reconfiguration. The examples show that their applicability strongly depends on the physical constraints of the considered system and the specification. Moreover, the examples present situations where the reconfigurability condition is not satisfied.

The reconfigurability theorem shows that no reconfiguration can be achieved for state internal failures where the system enters a state z_ε which was not modeled. In this case, no reconfigured controller exists even though the supercontroller has a nonzero redundancy degree. The main reason is the safe fulfillment of the specification which is no longer guaranteed. In the example of the manufacturing cell given in Chapter 7, every position in the space between, e.g., the deposit and the cooling block would be a failure state z_ε if the horizontal handler should unexpectedly stop there. In such a case, the automated control reconfiguration method presented here is not applicable. Instead, manual correcting actions are necessary.

Application examples with different grades of complexity have shown that the method is both manually and automatically applicable with an adequate processing device. The complexity analysis of the control design and the reconfiguration algorithm leads to polynomial complexity and, therefore, demonstrates the solvability of each problem.

8.2. Discussion and open problems

The literature reviews show that fault-tolerant control problems caught the interest of scientists a long time ago and still continues to be an area of interest. A selection of problems for future research is presented now.

Control loop autonomy. The control design problem is subject to ongoing research for several classes of discrete-event systems. The framework considered in this thesis is autonomous in the sense that no external influence is foreseen. In classic control theory of continuous systems, a reference signal is given to the controller which sends a control output signal to the plant until the error between the reference and the measurement vanishes. This concept needs to be adopted into discrete-event theory.

Another motivation is given by processes where a memory register is required, e.g., to check if some required products were processed in the specified order. In such a case, the model of the memory component, e.g., a shift register exponentially increases the complexity of the global model even for a small memory size. Therefore, it is more suitable to outsource the memory component from the plant model to an external unit which feeds the plant with reference values or set points.

An idea would be to declare the "reference" of the discrete-event controller as a target state to reach. The internal state sequence of the controller should be computed dynamically in a way that the controller stops sending events to the plant once the target state is reached. This problem can be solved by means of the symbolic adjacency matrices introduced in Section 2.2.3.

Time consideration. A reasonable model extension could be to include the notion of time in the models in order to capture some time issues which are relevant to real systems with unavoidable delays e.g. due to the diagnosis operation and the post-reconfiguration performance analysis. A corresponding controller synthesis method is proposed in [35].

Trade-off between on-line and off-line reconfiguration. The on-line reconfiguration can be used to deal with intermittent faults because the reconfiguration is applied locally and iteratively where the fault occurs. For permanent faults, the off-line reconfiguration method is well suited because it achieves a global reconfiguration of the controller so that any future effect of the fault is counter-acted in advance. However, off-line control reconfiguration might be expensive in terms of resources required for a successful reconfiguration. Thus, it is worth questioning how to combine the less demanding on-line reconfiguration with the global corrective actions of the off-line reconfiguration. In a sense, an extension would be to let the on-line reconfiguration "act locally" but "think globally". The reconfiguration algorithm should forecast the transitions where, e.g., a broken actuator might be needed again in the future and avoid its activation by the controller instead of letting the same fault being diagnosed again. To achieve this, it would be crucial to synchronize the pro-active actions of the reconfiguration with the diagnosis to avoid false or missing alarms. Obviously, a bidirectional interaction between the diagnosis unit and the reconfiguration unit would be necessary.

Fault-tolerant control of I/O automata networks. The fault-tolerant control framework presented in this thesis is based on monolithic models of the plant and the controller. The example of the manufacturing cell showed that it is often hard to build a monolithic model as an I/O automaton due to the problem of state space explosion. The complexity reduction achieved by a component-oriented modeling of the plant in terms of I/O automata networks leads to other questions concerning the appropriate architecture for nominal control and control reconfiguration. The results of this thesis hold for a centralized control architecture as in the case of the pilot manufacturing cell (Chapter 7). Diagnostic methods for I/O automata network were studied in [66, 139]. A new topic of interest would be to investigate under which conditions a modularly built control reconfiguration scheme could reach or even exceed the performance of a centralized control reconfiguration scheme.

Bibliography

Contributions by the author

Publications

[1] Y. Nke and J. Lunze. Fault-tolerant control of nondeterministic input/output automata subject to actuator faults. In *10th International Workshop on Discrete Event Systems (WODES'10)*, pages 360–365, Berlin, Sep. 2010.

[2] Y. Nke and J. Lunze. Control design for nondeterministic input/output automata. In *Proceedings of the 18th IFAC Congress*, pages 6994–6999, Milan, 2011.

[3] Y. Nke and J. Lunze. A fault modeling approach for input/output automata. In *Proceedings of the 18th IFAC Congress*, pages 8657–8662, Milan, 2011.

[4] Y. Nke and J. Lunze. Online control reconfiguration for a faulty manufacturing process. In *3rd International Workshop on Dependable Control of Discrete Systems (DCDS'11)*, pages 21–26, Saarbrücken, 2011.

[5] Y. Nke and J. Lunze. Control reconfiguration based on unfolding of input/output automata. In *Preprints of IFAC Symposium on Fault Detection Supervision and Safety for Technical Processes (SAFEPROCESS'12)*, pages 866–873, Mexico City, 2012.

[6] Y. Nke and J. Lunze. Systematischer entwurf fehlertoleranter steuerungen. In *Fachtagung zum Entwurf komplexer Automatisierungssysteme (EKA'12)*, Magdeburg, 2012.

[7] Y. Nke and J. Lunze. Systematischer entwurf fehlertoleranter steuerungen. *Automatisierungstechnik*, 61(2):122–130, Feb. 2013.

[8] Y. Nke and J. Lunze. On-line control reconfiguration for manufacturing systems. In J. Campos, C. Seatzu, and X. Xie, editors, *Formal Methods in Manufacturing*. CRC Press/Taylor and Francis, 2013. To appear.

[9] Y. Nke and J. Lunze. Control design for nondeterministic input/output automata. *European Journal of Control*. In preparation.

[10] Y. Nke, S. Drüppel, and J. Lunze. Direct feedback in asynchronous networks of input-output automata. In *Proceedings of the 10th European Control Conference (ECC'09)*, pages 2608–2613, Budapest, 2009.

Technical reports

[11] Y. Nke. http://www.atp.rub.de/ftcdes, Dec. 2012.

[12] Y. Nke. Entwicklung, implementierung und erprobung eines kompositions- und simulationsverfahrensfür automatennetze. Master's thesis, Ruhr-Universität Bochum, Bochum, Mar. 2008.

[13] Y. Nke. Fault-tolerant control of discrete event systems: problems and approaches. Technical report, Ruhr Universität Bochum, Lehrstuhl für Automatisierungstechnik und Prozessinformatik, 2009.

[14] Y. Nke. Control design of nondeterministic input/output automata. Technical report, Lehrstuhl für Automatisierungstechnik und Prozessinformatik, Ruhr-Universität, Bochum, Apr. 2010.

[15] Y. Nke. Formal design of deterministic controllers for input-/output automata with redundancies. Technical report, Lehrstuhl für Automatisierungstechnik und Prozessinformatik, Ruhr-Universität, Bochum, 2012.

[16] Y. Nke. Idefics - release 1.0 - an implementation of discrete-event fault tolerance in control systems in matlab/simulink. Technical report, Lehrstuhl für Automatisierungstechnik und Prozessinformatik, Ruhr-Universität, Bochum, 2013.

Supervised theses

[17] I.G. Altiok. Implementierung einer benutzeroberfläche zur simulation von eingangs/ausgangs-automaten. Bachelor thesis, Ruhr-Universität Bochum, Bochum, Nov. 2011.

[18] S. Bolat. Erprobung eines steuerungsentwurfsverfahrens für e/a-automatem am beispiel einer fertigungsanlage. Bachelor thesis, Ruhr-Universität Bochum, Bochum, Mar. 2011.

[19] J. Eisenburger. Komponentenorientierte modellbildung des dreitanksystems mit hilfe von eigangs/ausgangs-automaten. Bachelor thesis, Ruhr-Universität Bochum, Bochum, Aug. 2011.

[20] F. Gerecht. Ereignisdiskrete fehlertolerante steuerung einer fertigungsanlage. Diploma thesis, Ruhr-Universität, Lehrstuhl für Automatisierungstechnik und Prozessinformatik, Bochum, May 2012.

[21] P. Hohoff. Steuerungsentwurf eines virtuellen fertigungsprozesses anhand von eingangs/ausgangs-automaten. Bachelor thesis, Ruhr-Universität Bochum, Bochum, Mar. 2012.

[22] D. Husslein. Anwendung eines formalismus zum ereignisdiskreten steuerungsentwurf. Bachelor thesis, Ruhr-Universität Bochum, Bochum, Jul. 2011.

[23] L.d.G.P. López. Investigation of a faulty manufacturing process. Bachelor thesis, Ruhr-Universität, Lehrstuhl für Automatisierungstechnik und Prozessinformatik, Bochum, Oct. 2010.

[24] M. Schmidt. Erprobung und implementierung einer methode zur fehlertoleranten steuerung eines ereignisdiskreten verfahrenstechnischen prozesses. Bachelor thesis, Ruhr-Universität, Lehrstuhl für Automatisierungstechnik und Prozessinformatik, Bochum, Apr. 2010.

[25] P. Weizinger. Entwicklung und erprobung eines simulationsverfahrens zur rekonfiguration ereignisdiskreter systeme. Diploma thesis, Ruhr-Universität, Lehrstuhl für Automatisierungstechnik und Prozessinformatik, Bochum, Jul. 2011.

General literature

[26] *Proceedings of a Symposium on Large-Scale Digital Calculating Machinery: 7 - 10 Jan. 1947*. Charles Babbage Institute Reprint Series for the History of Computing. Harvard University Press, 1985.

[27] T. Aardenne-Ehrenfest and N.G. de Bruijn. Circuits and trees in oriented linear graphs. In I. Gessel and G.-C. Rota, editors, *Classic Papers in Combinatorics*, Modern Birkhäuser Classics, pages 149–163. Birkhäuser Boston, 1987.

[28] S.B. Akers. Binary decision diagrams. *IEEE Transactions on Computers*, C-27(6): 509–516, Jun. 1978.

[29] K. Akesson, M. Fabian, H. Flordal, and A. Vahidi. Supremica - a tool for verification and synthesis of discrete event supervisors. In *Proceedings of the 11th Mediterranean Conference on Control and Automation*, Rhodos, 2003.

[30] B. Alpern and F.B. Schneider. Recognizing safety and liveness. *Distributed Computing*, 2:117–126, 1987.

[31] K. Andersson, B. Lennartson, and M. Fabian. Synthesis of restart states for manufacturing cell controllers. In *2nd Ifac Workshop on Dependable Control of Discrete Systems*, pages 303–308, Bari, 2009.

[32] M. Arbib. Tolerance automata. *Kybernetika*, 3(3):223–233, 1967.

[33] A. Arora and S.S. Kulkarni. Component based design of multitolerant systems. *IEEE Transactions on Software Engineering*, 24(1):63–78, Jan. 1998.

[34] A. Arora and S.S. Kulkarni. Detectors and correctors: a theory of fault-tolerance components. In *Proceedings of the 18th International Conference on Distributed Computing Systems*, pages 436–443, May 1998.

[35] E. Asarin, O. Maler, and A. Pnueli. Symbolic controller synthesis for discrete and timed systems. In *Hybrid Systems II*, pages 1–20. Springer–Verlag, 1995.

[36] A. Avižienis. Fault-tolerance: The survival attribute of digital systems. *Proceedings of the IEEE*, 66(10):1109–1125, Oct. 1978.

[37] A. Avizienis, J.-C. Laprie, B. Randell, and C. Landwehr. Basic concepts and taxonomy of dependable and secure computing. *IEEE Transactions on Dependable and Secure Computing*, 1(1):11–33, Jan. 2004.

[38] F. Baccelli, G. Cohen, G. J. Olsder, and J.-P. Quadrat. *Synchronization and linearity*. Wiley and Sons, 2001.

[39] C. Baier and J.-P. Katoen. *Principles of model checking*. MIT-Press, Massachusetts, 2008.

[40] T. Bak, R. Wisniewski, and M. Blanke. Autonomous attitude determination and control system for the orsted satellite. *Proceedings of the IEEE Aerospace Applications Conference*, 2:173–186, Feb. 1996.

[41] D. Balageas, C-P. Fritzen, and Güemes, editors. *Structural Health Monitoring*. ISTE, 2006.

[42] S. Balemi, G.J. Hoffmann, P. Gyugyi, H. Wong-Toi, and G.F. Franklin. Supervisory control of a rapid thermal multiprocessor. *IEEE Transactions on Automatic Control*, 38(7):1040–1059, Jul. 1993.

[43] G. Barrett and S. Lafortune. Bisimulation, the supervisory control problem and strong model matching for finite state machines. *Discrete Event Dynamic Systems*, 8:377–429, 1998.

[44] N. Bauer, S. Engell, R. Huuck, S. Lohmann, B. Lukoschus, M. Remelhe, and O. Stursberg. Verification of plc programs given as sequential function charts. In *Integration of Software Specification Techniques for Applications in Engineering*, volume 3147 of *Lecture Notes in Computer Science*, pages 517–540. Springer Berlin / Heidelberg, 2004.

[45] E. Benoit, M-P. Huget, P. Moreaux, and O. Passalacqua. Reconfiguration of a distributed information fusion system. In *2nd Ifac Workshop on Dependable Control of Discrete Systems*, pages 309–314, Bari, 2009.

[46] A. Benveniste, E. Fabre, S. Haar, and C. Jard. Diagnosis of asynchronous discrete event systems, a net unfolding approach. *IEEE Transactions on Automatic Control*, 48(5):714–727, May 2003.

[47] L.T. Berger and K. Iniewski. *Smart Grid: Applications, Communications, and Security*. Wiley, New Jersey, 2012.

[48] M. Blanke and R.J. Patton. Industrial actuator benchmark for fault detection and isolation. *Control Engineering Practice*, 3(12):1727–1730, 1995.

[49] M. Blanke, M. Kinnaert, J. Lunze, and M. Staroswiecki. *Diagnosis and fault-tolerant control*. Springer Verlag, Heidelberg, 2nd. edition, 2006.

[50] D.W. Bradley and A.M. Tyrell. Immunotrics: Hardware fault tolerance inspired by the immune system. In *Proceedings of Third International Conference on Evolvable Systems (ICES 2000)*, pages 11–20, Apr. 2000.

[51] B. Braun. Fcpre: Extending the arora-kulkarni method of automatic addition of fault-tolerance. *ares*, 0:967–974, 2007.

[52] E. Brinksma and A. Mader. Model checking embedded system designs. In *6th Int. Workshop on Discrete Event Systems*, pages 151–158, Zaragoza, Oct. 2002.

[53] R.E. Bryant. Graph-based algorithms for boolean function manipulation. *IEEE Transactions on Computers*, C-35(8):677–691, Aug. 1986.

[54] M. Cantarelli and J.-M. Roussel. Reactive control system design using the supervisory control theory: evaluation of possibilities and limits. In *Proceedings of the 9th International Workshop on Discrete Event Systems*, pages 200–205, Göteborg, 2008.

[55] L.K. Carvalho, J.C. Basilio, and M.V. Moreira. Robust diagnosability of discrete event systems subject to intermittent sensor failures. In *Workshop on Discrete Event Systems*, Berlin, 2010.

[56] C.G. Cassandras and S. Lafortune. *Introduction to discrete event systems*. Springer Science and Business Media, LLC, New York, 2008.

[57] D.A. Castanon. Efficient algorithms for finding the k best paths through a trellis. *Aerospace and Electronic Systems, IEEE Transactions on*, 26(2):405–410, Mar. 1990.

[58] E. M. Clarke, O. Grumberg, and D. A. Peled. *Model checking*. The MIT Press, London, 2nd edition, 2000.

[59] L.B. Clavijo, J.C. Basilio, and L.K. Carvalho. Deslab: A scientific computing program for analysis and synthesis of discrete-event systems. In *Proceedings of the 11th International Workshop on Discrete Event Systems*, pages 349–355, Guadalajara, 2012.

[60] L. Console and O. Dressier. Model-based diagnosis in the real world: lessons learned and challenges remaining. In *Proceedings of the 16th international joint conference on Artificial intelligence*, pages 1393–1400, 1999.

[61] M. Dal Cin. Verifying fault-tolerant behavior of state machines. In *Proceedings of the 2nd High-Assurance Systems Engineering Workshop*, pages 94–99, Washington, Aug. 1997.

[62] R. Davis and W. Hamscher. Model-based reasoning: troubleshooting. pages 297–346, 1988.

[63] J. Desel, H.-M. Hanisch, G. Juhás, R. Lorenz, and C. Neumair. A guide to modelling and control with modules of signal nets. In *Integration of Software Specification Techniques for Applications in Engineering*, volume 3147 of *Lecture Notes in Computer Science*, pages 270–300. Springer Berlin / Heidelberg, 2004.

[64] M.D. DiBenedetto, A. Saldanha, and A. Sangiovanni-Vincentelli. Strong model matching for finite state machines with non-deterministic reference model. In *Proceedings of the 34th IEEE Conference on Decision and Control*, volume 1, pages 422–426, 1995.

[65] S.X. Ding. A survey of fault-tolerant networked control system design. In *Preprints of IFAC Symposium on Fault Detection Supervision and Safety for Technical Processes*, pages 874–885, Mexico City, 2012.

[66] S. Drüppel. *Modeling and partially coordinated diagnosis of asynchronous discrete-event systems*. PhD thesis, Ruhr-Universität Bochum, Hannover, 2012.

[67] E. Dumitrescu, A. Girault, H. Marchand, and E. Rutten. Optimal discrete controller synthesis for modeling fault-tolerant distributed systems. In *IFAC Workshop on Dependable Control of Discrete-Event Systems*, pages 23–28, Cachan, 2007.

[68] A. Ebnenasir. *Automatic synthesis of fault-tolerance*. PhD thesis, East Lansing, 2005. AAI3171453.

[69] J. Ekanayake, K. Liyanage, J. Wu, A. Yokoyama, and N. Jenkins. *Smart Grid: Technology and Applications*. Wiley, USA, 2012.

[70] J.G. Eldredge and B.L. Hutchings. Density enhancement of a neural network using fpgas and run-time reconfiguration. In *Proceedings of the IEEE Workshop on FPGAs for Custom Computing Machines*, pages 180–188, Apr. 1994.

[71] J.G. Eldredge and B.L. Hutchings. Rrann: a hardware implementation of the back-propagation algorithm using reconfigurable fpgas. In *IEEE International Conference on Neural Networks*, pages 2097–2102, 1994.

[72] J.G. Eldredge and B.L. Hutchings. Rrann: the run-time reconfiguration artificial neural network. In *Proceedings of the IEEE Custom Integrated Circuits Conference*, pages 77–80, May 1994.

[73] C.R. Farrar and N.A.J Lieven. Damage prognosis: the future of structural health monitoring. *Philosophical Transactions of the Royal Society A: Mathematical, Physical and Engineering Sciences*, 365(1851):623–632, 2007.

[74] D. Floreano and C. Mattiussi. *Bio-Inspired Artificial Intelligence*. MIT Press, USA, 2008.

[75] D. Floreano, N. Schoeni, G. Caprari, and J. Blynel. Evolutionary bits'n'spikes. In *Proceedings of the Eight International Conference On Artificial Life*, Cambridge, 2002.

[76] C. Frei, F. Kraus, and M. Blanke. Recoverability viewed as a system property. In *European Control Conference*, Karlsruhe, 1999.

[77] H.E. Garcia, A. Ray, and R.M. Edwards. A reconfigurable hybrid supervisory system for process control. In *Proceedings of the 33rd IEEE Conference on Decision and Control*, volume 3, pages 3131–3136, Dec. 1994.

[78] F. Gärtner and A. Jhumka. Automating the addition of fail-safe fault-tolerance: Beyond fusion-closed specifications. In *Formal Techniques, Modelling and Analysis of Timed and Fault-Tolerant Systems*, volume 3253 of *Lecture Notes in Computer Science*, pages 19–24. Springer Berlin / Heidelberg, 2004.

[79] X. Geng and J. Hammer. Input/output control of asynchronous sequential machines. *IEEE Transactions on Automatic Control*, 50(12):1956–1970, Dec. 2005.

[80] A. Girault and E. Rutten. Automating the addition of fault tolerance with discrete controller synthesis. *Formal Methods in System Design*, 35(2):190–225, 2009.

[81] V.-M. Glushkov. The abstract theory of automata. *Russian Mathematical Surveys*, 16 (5):1–53, Oct. 1961. translation from Uspekhi Matematicheskikh Nauk 16:5 (1961), pp. 3–62.

[82] C.H. Golaszewski and P.J. Ramadge. Control of discrete event processes with forced events. In *26th IEEE Conference on Decision and Control*, pages 247–251, Dec. 1987.

[83] V. Gourcuff, O. De Smet, and J.M. Faure. Efficient representation for formal verification of plc programs. In *8th International Workshop on Discrete Event Systems*, pages 182–187, 2006.

[84] J Grainger. Cracking the orthographic code: An introduction. *Language and Cognitive Processes*, 23(1):1–35, 2008.

[85] R.P. Grimaldi. *Discrete and Combinatorial Mathematics*. Addison Wesley Longman, 1999.

[86] J. L. Gross and J. Yellen. *Handbook of graph theory*. CRC Press, USA, 2004.

[87] J. Hammer. On some control problems in molecular biology. In *Proceedings of the 33rd IEEE Conference on Decision and Control*, Lake Buena Vista, 1994.

[88] J. Hammer. On the modeling and control of biological signaling chains. In *Proceedings of the 34th IEEE Conference on Decision and Control*, New Orleans, 1995.

[89] J. Hammer. On the corrective control of sequential machines. *International Journal of Control*, 65(2):249–276, 1996.

[90] J. Hammer. On the control of incompletely described sequential machines. *International Journal of Control*, 63(6):1005–1028, 1996.

[91] J. Hammer. On the control of sequential machines with disturbances. *International Journal of Control*, 67(3):307–331, 1997.

[92] F. Harary. *Graph theory*. Perseus Book Publishing, L.L.C., Boulder, 1969.

[93] C.A.R. Hoare. Communicating sequential processes. *Communications of the ACM*, 21(8):666–677, Aug. 1978.

[94] C.A.R. Hoare. *Communicating sequential processes*. Prentice-Hall, 1985.

[95] G. Hoffmann and H. Wong-Toi. Symbolic synthesis of supervisory controllers. In *American Control Conference*, pages 2789–2793, Jun. 1992.

[96] L.E. Holloway, B.H. Krogh, and A. Giua. A survey of petri net methods for controlled discrete event systems. *Discrete Event Dynamic Systems*, 7:151–190, 1997.

[97] J.E. Hopcroft, R. Motwani, and J.D. Ullman. *Introduction to automata theory, languages, and computation*. Addison-Wesley, New York, 2001.

[98] J. Huang and R. Kumar. An optimal directed control framework for discrete event systems. *IEEE Transactions on Systems, Man and Cybernetics, Part A: Systems and Humans*, 37(5):780–791, Sep. 2007.

[99] J. Huang and R. Kumar. Optimal nonblocking directed control of discrete event systems. *IEEE Transactions on Automatic Control*, 53(7):1592–1603, Aug. 2008.

[100] D.J. Inman, C.R. Farrar, V.L. Junior, and V.S. Junior. *Damage Prognosis: For Aerospace, Civil and Mechanical Systems*, pages 1–19. John Wiley & Sons, 2005.

[101] Technical Committee 65B International Electrotechnical Commission (IEC). *IEC 61131-3 - Programmable controllers - Part 3: Programming languages*, 2nd. edition, 2003.

[102] R.M. Jensen, R.E. Bryant, and M.M. Veloso. Seta*: An efficient bdd-based heuristic search algorithm. In *AAAI/IAAI'02*, pages 668–673, 2002.

[103] Rune M. Jensen. Des controller synthesis and fault tolerant control - a survey of recent advances. Technical Report TR-2003-40, IT University of Copenhagen, Copenhagen, 2003.

[104] G. Juhás, R. Lorenz, and C. Neumair. Modelling and control with modules of signal nets. In *Lectures on Concurrency and Petri Nets*, volume 3098, pages 1–22. Springer Berlin / Heidelberg, 2004.

[105] G. Kalyon, T. Le Gall, H. Marchand, and T. Massart. Symbolic supervisory control of infinite transition systems under partial observation using abstract interpretation. *Discrete Event Dynamic Systems*, 22:121–161, 2012.

[106] S. Karras. *Systematischer modellgestützter Entwurf von Steuerungen für Fertigungssysteme*. PhD thesis, Martin-Luther-Universität Halle-Wittenberg, Halle (Saale), 2007.

[107] Z. Kohavi and N.K. Jha. *Switching and finite automata theory*. Cambridge Univ. Press, New York, 3rd edition, 2009.

[108] A. Kulkarni. *Component Based Design Of Fault-Tolerance*. PhD thesis, 1999.

[109] S. Kulkarni and A. Arora. Automating the addition of fault-tolerance. In *Formal Techniques in Real-Time and Fault-Tolerant Systems*, volume 1926 of *Lecture Notes in Computer Science*, pages 339–359. Springer Berlin / Heidelberg, 2000.

[110] R. Kumar and S. Takai. A framework for control-reconfiguration following fault-detection in discrete event systems. In *Preprints of IFAC Symposium on Fault Detection Supervision and Safety for Technical Processes*, pages 848–853, Mexico City, 2012.

[111] S. Lamperiere-Couffin and J.J. Lesage. Formal verification of the sequential part of plc programs. In *5th Workshop on Discrete Event Systems*, pages 247–254, Ghent, Aug. 2000.

[112] G. Lamperti and M. Zanella. *Diagnosis of active systems*. Kluwer Academic Publishers, Dordrecht, 2003.

[113] L. Lamport. Proving the correctness of multiprocess programs. *IEEE Trans. Softw. Eng.*, SE-3:125–143, Mar. 1977.

[114] L. Lamport, R. Shostak, and M. Pease. The byzantine generals problem. *ACM Transactions on Programming Languages and Systems*, 4(3):382–401, Jul. 1982.

[115] C.G. Langton. Self-reproduction in cellular automata. *Physica D: Nonlinear Phenomena*, 10(1-2):135–144, 1984.

[116] A. Lavoisier. *Traité élémentaire de chimie (Elementary Treatise of Chemistry)*. 1789.

[117] C.M. Lee. Graph-based algorithms for boolean function manipulation. *Bell System Technical Journal*, 38:985–999, 1959.

[118] E.A. Lee and S.A. Seshia. *Introduction to Embedded Systems*. LuLu, USA, 2011.

[119] P.A. Lee and Anderson T. *Fault tolerance - Principles and Practice*, volume 3. Springer-Verlag Wien New York, 1990.

[120] S-H. Lee and C-S. Wu. A k-best paths algorithm for highly reliable communication networks. *IEICE Trans. Communications*, E82-B(4):586–590, Apr. 1999.

[121] G.-L. Li. State feedback control of vector discrete event systems with forced events. In *International Conference on Machine Learning and Cybernetics*, pages 469–471, Aug. 2007.

[122] Y.-T.S. Li and S. Malik. Performance analysis of embedded software using implicit path enumeration. *IEEE Transactions on Computer-Aided Design of Integrated Circuits and Systems*, 16(12):1477–1487, Dec. 1997.

[123] F. Lin and W.M. Wonham. On observability of discrete-event systems. *Information sciences*, 44(3):173–198, Apr. 1988.

[124] O. Ljungkrantz, K. Akesson, J. Richardsson, and K. Andersson. Implementing a control system framework for automatic generation of manufacturing cell controllers. In *Proceedings of the IEEE International Conference on Robotics and Automation*, pages 674–679, Roma, 2007.

[125] A. Lüder. *Formaler Steuerungsentwurf mit modularen diskreten Verhaltensmodellen*. PhD thesis, Martin-Luther-Universität Halle-Wittenberg, Halle (Saale), 2007.

[126] J. Lunze. *Ereignisdiskrete Systeme*. Oldenbourg Verlag, München Wien, 2006.

[127] J. Lunze. Relations between networks of standard automata and networks of i/o automata. In *Proceedings of the 9th International Workshop on Discrete Event Systems*, pages 425–430, Göteborg, 2008.

[128] J. Lunze and J.H. Richter. Reconfigurable fault-tolerant control: A tutorial introduction. *European Journal of Control*, 14(5):359–386, 2008.

[129] J. Lunze and J. Schröder. Sensor and actuator fault diagnosis of systems with discrete inputs and outputs. *IEEE Transactions on Systems, Man, and Cybernetics, Part B*, 34(2):1096–1107, Apr. 2004.

[130] Y. Ma, Q. Yu, and I. Cohen. Multiple hypothesis target tracking using merge and split of graph's nodes. In *Advances in Visual Computing*, volume 4291 of *Lecture Notes in Computer Science*, pages 783–792. Springer Berlin / Heidelberg, 2006.

[131] H. Marchand, P. Bournai, M. Le Borgne, and P. Le Guernic. Synthesis of discrete-event controllers based on the signal environment. *Discrete Event Dynamic Systems*, 10:325–346, 2000.

[132] C. Maul. Implementierung und erprobung eines diagnoseverfahrens für automatennetze. Bachelor thesis, Ruhr-Universität Bochum, Lehrstuhl für Automatisierungstechnik und Prozessinformatik, 2008.

[133] R. Mc Naughton and H. Yamada. Regular expressions and state graphs for automata. *Electronic Computers, IRE Transactions on*, 9(1):39–47, 1960.

[134] N. Mechbal and E.G.O. Nóbrega. Damage tolerant active control: Concept and state of art. In *Preprints of IFAC Symposium on Fault Detection Supervision and Safety for Technical Processes*, pages 63–71, Mexico City, 2012.

[135] J. Momoh. *Smart Grid: Fundamentals of Design and Analysis*. Wiley, USA, 2012.

[136] T.E. Murphy, X. Geng, and J. Hammer. Controlling races in asynchronous sequential machines. In *Proceedings of the 15th IFAC Congress*, Barcelona, 2002.

[137] T.E. Murphy, X. Geng, and J. Hammer. On the control of asynchronous machines with races. *IEEE Transactions on Automatic Control*, 48(6):1073–1081, Jun. 2003.

[138] G. Mušič, D. Matko, and B. Zupančič. Modelling, synthesis, and simulation of supervisory process control systems. *Mathematical and Computer Modelling of Dynamical Systems*, 6(2):169–189, 2000.

[139] J. Neidig. *An automata theoretic approach to modular diagnosis of discrete-event systems*. PhD thesis, Ruhr-Universität Bochum, Nuernberg, Apr. 2007.

[140] S.D. Nikolopoulos, A. Pitsillides, and D. Tipper. Addressing network survivability issues by finding the k-best paths through a trellis graph. In *Sixteenth Annual Joint Conference of the IEEE Computer and Communications Societies. Proceedings IEEE*, volume 1, pages 370–377, Kobe, Apr. 1997.

[141] M.T. Omran, J-R. Sack, and H. Zarrabi-Zadeh. Finding paths with minimum shared edges. In *17th Annual International Computing and Combinatorics Conference*, Dallas, Aug. 2011.

[142] F. Ortmeier, M. Güdemann, and W. Reif. Formal failure models. In *Proceedings of the 1st IFAC Workshop on Dependable Control of Discrete Systems*, 2007.

[143] A.K. Pandey, M. Biswas, and M.M. Samman. Damage detection from changes in curvature mode shapes. *Journal of Sound and Vibration*, 145(2):321–332, 1991.

[144] A. Paoli, M. Sartini, and S. Lafortune. A fault tolerant architecture for supervisory control of discrete event systems. In *Proceedings of the 17th IFAC World Congress*, pages 6542–6547, Seoul, Jul. 2008. IFAC.

[145] Y. Papadopoulos and J.A. McDermid. Automated safety monitoring: A review and classification of methods. *International Journal of Condition Monitoring and Diagnostic Engineering Management*, 4(4):14–32, Oct. 2001.

[146] S.-J. Park and K.-H. Cho. Supervisory control for fault-tolerant scheduling of real-time multiprocessor systems with aperiodic tasks. *International Journal of Control*, 82(2):217–227, Feb. 2009.

[147] S.-J. Park and J.-T. Lim. Fault-tolerant robust supervisor for des with model uncertainty and its application to a workcell. *IEEE Transactions on robotics and automation*, 15(2):386–391, Apr. 1999.

[148] J. Peng and J. Hammer. Input/output control of asynchronous sequential machines with races. *International Journal of Control*, 83(1):125–144, 2010.

[149] S. Perk, T. Moor, and K. Schmidt. Controller synthesis for an i/o-based hierarchical system architecture. In *Proceedings of the 9th International Workshop on Discrete Event Systems*, pages 474–479, Göteborg, 2008.

[150] M. Petreczky, R.J.M. Theunissen, R. Su, D.A. van Beek, J.H. van Schuppen, and J.E. Rooda. Control of input/output discrete-event systems. In *Proceedings of the 10th European Control Conference*, Budapest, 2009.

[151] C. Picardi, L. Console, and D. Theseider Dupré. Model synthesis for model-based fault analysis. In *15th International Workshop on Principles of Diagnosis (DX'04)*, Carcassonne, Jun. 2004.

[152] L. E. Pinzon, H.-M. Hanisch, M. A. Jafari, and T. Boucher. A comparative study of synthesis methods for discrete event controllers. *Formal Methods in System Design*, 15(2):123–167, Sep. 1999.

[153] I. Pomeranz and S.M. Reddy. On achieving a complete fault coverage for sequential machines using the transition fault model. In *28th ACM/IEEE Design Automation Conference*, pages 341–346, Jun. 1991.

[154] G. Provan and Y-L. Chen. Model-based fault tolerant control reconfiguration for discrete event systems. In *Proceedings of the 2000 IEEE International Conference on Control Applications*, pages 473–478, Anchorage, Sep. 2000.

[155] H.A. Rademacher. On the accumulation of errors in processes of integration on high-speed calculating machines. In *Charles Babbage Institute Reprint Series for the History of Computing*, volume 7, pages 176–177, Cambridge, MA, Jan. 1947.

[156] P.J. Ramadge and W.M. Wonham. The control of discrete event systems. *Proceedings of the IEEE*, 77(1):81–98, Jan. 1989.

[157] M. Rausch and B.H. Krogh. Formal verification of plc programs. In *Proceedings of the American Control Conference*, volume 1, pages 234–238, Jun. 1998.

[158] G. Rawlinson. The significance of letter position in word recognition. Unpublished PhD Thesis, Psychology Department, University of Nottingham, 1976.

[159] G. Rawlinson. Reibadailty. *New Scientist, Letters*, 162:55, May 1999.

[160] K. Rayner, S. J. White, R. L. Johnson, and S. P. Liversedge. Raeding wrods with jubmled lettres: There is a cost. *Psychological Science*, 192–193:55, Mar. 2006.

[161] Realgames. http://www.realgames.pt/, Nov. 2011.

[162] R. Reiter. A theory of diagnosis from first principles. *Artificial Intelligence*, 32(1): 57–95, 1987.

[163] L. Ricker, S. Lafortune, and S. Genc. Desuma: A tool integrating giddes and umdes. In *Proceedings of the 8th International Workshop on Discrete-Event Systems*, 2006.

[164] D. Roggen, S. Hofmann, Y. Thoma, and D. Floreano. Hardware spiking neural network with run-time reconfigurable connectivity in an autonomous robot. In *Proceedings of the NASA/DoD Conference on Evolvable Hardware*, pages 189–198, Jul. 2003.

[165] K.R. Rohloff. Sensor failure tolerant supervisory control. In *Proceedings of 44th IEEE Conference on Decision and Control, and the European Control Conference*, Seville, 2005.

[166] J.-M. Roussel and A. Giua. Designing dependable logic controllers using the supervisory control theory. In *Proceedings of the 16th IFAC Congress*, Prague, 2005.

[167] M. Sachenbacher and P. Struss. Automated qualitative domain abstraction. In *18th International Joint Conference on Artificial Intelligence*, pages 382–387, Acapulco, 2003.

[168] R.P.C. Sampaio, N.M.M. Maia, and J.M.M. Silva. Damage detection using the frequency-response-function curvature method. *Journal of Sound and Vibration*, 226(5):1029–1042, 1999.

[169] M. Sampath, R. Sengupta, S. Lafortune, K. Sinnamohideen, and D. Teneketzis. Diagnosability of discrete event systems. *IEEE Transactions on Automatic Control*, 40 (9):1555–1575, Sep. 1995.

[170] A. Sanchez, G. Rotstein, N. Alsop, and S. Macchietto. Synthesis and implementation of procedural controllers for event-driven operations. *AIChE Journal*, 45(8):1753–1775, 1999.

[171] M. Sayed Mouchaweh, A. Philippot, V. Carré-Ménétrier, and B. Riera. Fault diagnosis of discrete event systems using components fault-free models. In *20th International Workshop on Principles of Diagnosis*, Stockholm, Jun. 2009.

[172] S. Simani, C. Fantuzzi, and R.J. Patton. *Model-based fault diagnosis in dynamic system using identification techniques*. Springer Verlag, London, 2003.

[173] M. Staroswiecki, C. Commault, and J.-M. Dion. Component usefulness measures for fault tolerance evaluation. In *Preprints of IFAC Symposium on Fault Detection Supervision and Safety for Technical Processes*, pages 510–515, Mexico City, 2012.

[174] K. Stoy, D. Brandt, and D.J. Christensen. *Self-reconfigurable robots : an introduction*. Intelligent robotics and autonomous agents. Cambridge, Massachussets: MIT Press, London, 2010.

[175] M. Sugita. Functional analysis of chemical systems in vivo using a logical circuit equivalent. ii. the idea of a molecular automaton. *Journal of Theoretical Biology*, 4 (2):179–192, 1963.

[176] TCT. http://www.control.utoronto.ca/DES/, Nov. 2012.

[177] S. Tripakis. Fault diagnosis for timed automata. In *7th International Symposium on Formal Techniques in Real-Time and Fault-Tolerant Systems*, pages 205–224, London, 2002. Springer-Verlag.

[178] E. Tronci. Automatic synthesis of controllers from formal specifications. In *Proceedings of the Second IEEE International Conference on Formal Engineering Methods*, Washington, 1998.

[179] N. Venkatraman and J. Hammer. On the control of asynchronous sequential machines with infinite cycles. *International Journal of Control*, 79(7):764–785, 2006.

[180] N. Venkatraman and J. Hammer. Controllers of asynchronous sequential machines with infinite cycles. In *Proceedings of the 17th International Symposium on Mathematical Theory of Networks and Systems*, Kyoto, 2006.

[181] J. von Neumann. *Theory of Self-ReproducingAutomata*. University of Illinois Press, Illinois, 1966.

[182] G. A. Wainer and P. J. Mostermann. *Discrete-Event Modeling and Simulation*. CRC Press, USA, 2011.

[183] J.E. Walter, J.L. Welch, and N.M. Amato. Distributed reconfiguration of metamorphic robot chains. In *Proc. of ACM SIGACT-SIGOPS Symp. on Princ. of Dist. Comp.*, pages 171–180, 2000.

[184] J.E. Walter, J.L. Welch, and N.M. Amato. Distributed reconfiguration of metamorphic robot chains. *Distributed Computing*, 17:171–189, 2004.

[185] M. W. Warner. Cost automata, optimality and tolerance. *Kybernetes: The International Journal of Systems and Cybernetics*, 11(1):47–51, 1982.

[186] P. Weber, C. Simon, D. Theilliol, and V. Puig. Fault-tolerant control design for over-actuated system conditioned by reliability: a drinking water network application. In *Preprints of IFAC Symposium on Fault Detection Supervision and Safety for Technical Processes*, pages 558–563, Mexico City, 2012.

[187] Q. Wen, R. Kumar, J. Huang, and H. Liu. Weakly fault-tolerant supervisory control of discrete event systems. In *Proceedings of the American Control Conference*, New York, 2007.

[188] A.W. Wundheiler. The necessity of error analysis in numerical computation. In *Charles Babbage Institute Reprint Series for the History of Computing*, volume 7, pages 83–87, Cambridge, MA, Jan. 1947.

[189] S. Xanthakis, S. Karapoulios, R. Pajot, and A. Rozz. Immune system and fault-tolerant computing. In *Artificial Evolution*, volume 1063 of *Lecture Notes in Computer Science*, pages 181–197. Springer Berlin / Heidelberg, 1996.

[190] F. Xue and D-Z. Zheng. Fault-tolerant supervisory for discrete event systems based on event observer. In *Proceedings (Part II) of the International Conference on Intelligent Computing*, pages 655–664, Hefei, Anhui, 2005.

[191] K. Yamalidou, J. Moody, M. Lemmon, and P. Antsaklis. Feedback control of petri nets based on place invariants. *Automatica*, 32(1):15–28, 1996.

[192] J.-M. Yang and J. Hammer. State feedback control of asynchronous sequential machines with adversarial inputs. *International Journal of Control*, 81(12):1910–1929, 2008.

[193] J.-M. Yang and J. Hammer. Counteracting the effects of adversarial inputs on asynchronous sequential machines. In *Proceedings of the 17th IFAC Congress*, Seoul, 2008.

[194] N. Yevtushenko, T. Villa, R. Brayton, A. Petrenko, and A. Sangiovanni-Vincentelli. Compositionally progressive solutions of synchronous fsm equations. *Discrete Event Dynamic Systems*, 18:51–89, 2008.

[195] S.Q. Zheng, B. Yang, M. Yang, and J. Wang. Finding minimum-cost paths with minimum sharability. In *26th IEEE International Conference on Computer Communications*, pages 1532–1540, Anchorage, May 2007.

Part III.

Appendices

A. Nomenclature

General notation

Description	Notation	Example
Scalars	small, italic	z, s
Vectors	small, bold, italic	$\boldsymbol{z}, \boldsymbol{s}$
Constant values	italic, with ‾	$\overline{Z}, \overline{\boldsymbol{A}}$
MATLAB names	teletype	`buildcontroller()`, `Np`, `zF`
Active set operator	Subscript a	$\mathcal{Z}_{ap}, \mathcal{W}_{ac}$
Plant related variables	Subscript p	$z_p, V_p, \mathcal{W}_{ap}$
Specification related variables	Subscript s	$w_s, \mathcal{V}_{as}$
Controller related variables	Subscript c	$v_c, \mathcal{N}_c, L_c$
Control loop related variables	Subscript l	$\mathcal{N}_l, w_l$
Fault related variables	Superscript f	L_p^f, v_p^f
Falsified signals	Subscript $\sharp$	$v_\sharp, w_\sharp, z_\sharp$
Signal failures	Subscript ε	$v_\varepsilon, w_\varepsilon, z_\varepsilon'$
Composition related variables	Superscript $\tilde{\ }$	$\tilde{\boldsymbol{G}}, \tilde{\mathcal{W}}_c(z_c, z_p)$
Reconfiguration related variables	Superscript r	$\mathcal{A}_c^r, w_c^r$
Nominal case related variables	Superscript n	w_c^n

Symbols

Symbol	Description
z'	Next state
z_0	Initial state

$\mathcal{Z}_0$	Set of possible initial states
z_F	Final state
$\mathcal{Z}_F$	Set of possible final states
v	Input event
w	Output event
f	Fault event
s	Interconnection input event
r	Interconnection output event
k	Step of a sequence
k_e	Final step of sequence
k_H	Final step for a reachability analysis
k_{ei}	Final state of the i-th sequences
$\boldsymbol{K_e}$	Set of horizons
$Z(0 \cdots k_e)$	Sequence of states with $k_e + 1$ elements
$V(0 \cdots k_e)$	Sequence input events with $k_e + 1$ elements
$W(0 \cdots k_e)$	Sequence output events with $k_e + 1$ elements
$Z(k)$	k-th element of a state sequence $Z(0 \cdots k_e)$
$V(k)$	k-th element of an input sequence $V(0 \cdots k_e)$
$W(k)$	k-th element of an output sequence $W(0 \cdots k_e)$
L	Characteristic function of an I/O automaton
λ	Characteristic function of an autonomous automaton
η	Characteristic function of I/O trellis automaton
$\mathcal{V}$	Set of inputs
$\mathcal{W}$	Set of outputs
$\mathcal{Z}$	Set of states
$\mathcal{F}$	Set of faults
$\mathcal{N}_S$	Interconnection inputs set
$\mathcal{N}_R$	Interconnection outputs set
ι	Index of a component in a network of extended I/O automata

μ	Maximal number of components in a network of extended I/O automata
i	Index of a control law $\mathcal{A}_c^{(i)}$
ν	Maximal number of control laws
F	Interconnection output function
G	State transition function
H	Output function
$\mathcal{S}$	Specification for an I/O automaton
$\mathcal{N}$	Nondeterministic I/O automaton
$\mathcal{A}$	Deterministic I/O automaton
$\mathcal{M}$	Autonomous I/O automaton
$\mathcal{H}$	I/O trellis automaton
$\boldsymbol{\mathcal{A}}$	Composed I/O automaton
$\boldsymbol{K}$	Coupling matrix of an I/O automata network
$\mathcal{AN}$	I/O automata network
ε	Empty symbol
$*$	"dontcare"-symbol
R	Cardinal number of the set of outputs
P	Cardinal number of the interconnection outputs set
N	Cardinal number of the set of states
M	Cardinal number of the set of inputs
T	Cardinal number of the interconnection inputs set
$\boldsymbol{A}^{k_H}$	Numeric adjacency matrix of an I/O automaton for k_H steps
$\overline{\boldsymbol{A}}$	Numeric adjacency matrix of an I/O automaton for the maximal horizon $k_H = N - 1$
$\boldsymbol{A}_z$	Symbolic state adjacency matrix
$\boldsymbol{A}_v$	Symbolic input adjacency matrix
$\boldsymbol{A}_w$	Symbolic output adjacency matrix
$\mathbb{A}_z$	Set of symbolic state adjacency matrices
$\mathbb{A}_v$	Set of symbolic input adjacency matrices
$\mathbb{A}_w$	Set of symbolic output adjacency matrices

$\boldsymbol{Z}(0 \cdots k_e)$	State sequence matrix of an I/O automata network
$\boldsymbol{V}(0 \cdots k_e)$	Input sequence matrix of an I/O automata network
$\boldsymbol{W}(0 \cdots k_e)$	Output sequence matrix of an I/O automata network
$\mathcal{Z}(0 \cdots \boldsymbol{K_e})$	Set of state sequences of an I/O automaton
κ	Maximal number of state sequences in $\mathcal{Z}(0 \cdots \boldsymbol{K_e})$
Err_v	Input-error relation
Err_w	Output-error relation
$Err_{z'}$	State-error relation
$\mathcal{I}$	Identity map
$\mathcal{C}$	Complexity of computattion

Operators

Symbol	Description
$\mathcal{S} \models X$	Specification $\mathcal{S}$ models the constraint X
$\wedge$	Boolean AND for two expressions
$\bigwedge$	Recursive Boolean AND for several expression
$\vee$	Boolean OR for two expression
$\bigvee$	Recursive Boolean OR for several expression
$\equiv$	Equivalence
$|\cdot|$	Cardinal number of a set
$\times$	Cartesian product
$\bigtimes$	Recursive Cartesian product of several sets
$\mathcal{V}_a(\cdot)$	Active input operator
$\mathcal{W}_a(\cdot)$	Active output operator
$\mathcal{Z}_a(\cdot)$	Active next state operator
$\mathrm{Spec}(\cdot)$	Specification operator
$\mathrm{Con}(\cdot)$	Control design operator
$\mathrm{BFS}(\cdot)$	Breadth-First-Search operator
$\Longrightarrow$	State sequence extraction from a specification $\mathcal{S}$

$od(z)$	Output degree of a state z
$td(z', z)$	Transition degree from state z to state z'
$E_v(\cdot)$	Input event falsification
$E_w(\cdot)$	Output event falsification
$E_{z'}(\cdot)$	Next state falsification
$(\cdot)^*$	Infinite repetition of $(\cdot)$
$df(\cdot)$	Degree of freedom of a specification
$rd_W(z)$	Output redundancy degree of state z
$rd_Z(z)$	State redundancy degree of state z
$RD(\mathcal{N})$	Redundancy degree of an automaton $\mathcal{N}$

B. Proofs

Proof of Lemma 2.1

Assume the nondeterministic I/O automaton $\mathcal{N}$ is W-deterministic and the negation of (2.47) holds, thus,

$$\neg(\forall\, (z, v) \in \mathcal{Z} \times \mathcal{V} : \mathcal{Z}_a(z, v) \neq \emptyset \Rightarrow |\mathcal{W}_a(z, v)| = 1) \tag{B.1}$$

$$\Leftrightarrow \quad \exists\, (z, v) \in \mathcal{Z} \times \mathcal{V} : \neg(\mathcal{Z}_a(z, v) \neq \emptyset \Rightarrow |\mathcal{W}_a(z, v)| = 1) \tag{B.2}$$

$$\Leftrightarrow \quad \exists\, (z, v) \in \mathcal{Z} \times \mathcal{V} : \neg(\neg(\mathcal{Z}_a(z, v) \neq \emptyset) \vee (|\mathcal{W}_a(z, v)| = 1)) \tag{B.3}$$

$$\Leftrightarrow \quad \exists\, (z, v) \in \mathcal{Z} \times \mathcal{V} : \neg(\mathcal{Z}_a(z, v) = \emptyset \vee |\mathcal{W}_a(z, v)| = 1) \tag{B.4}$$

$$\Leftrightarrow \quad \exists\, (z, v) \in \mathcal{Z} \times \mathcal{V} : \neg(\mathcal{Z}_a(z, v) = \emptyset) \wedge \neg(|\mathcal{W}_a(z, v)| = 1) \tag{B.5}$$

$$\Leftrightarrow \quad \exists\, (z, v) \in \mathcal{Z} \times \mathcal{V} : \mathcal{Z}_a(z, v) \neq \emptyset \wedge |\mathcal{W}_a(z, v)| \neq 1 \tag{B.6}$$

$$\Leftrightarrow \exists\, (z, v) \in \mathcal{Z} \times \mathcal{V} : \mathcal{Z}_a(z, v) \neq \emptyset \wedge (|\mathcal{W}_a(z, v)| < 1 \vee |\mathcal{W}_a(z, v)| > 1) \tag{B.7}$$

$$\Leftrightarrow \quad \exists\, (z, v) \in \mathcal{Z} \times \mathcal{V} :$$

$$\underbrace{(\mathcal{Z}_a(z, v) \neq \emptyset \wedge (|\mathcal{W}_a(z, v)| < 1)}_{(I)} \vee \underbrace{(\mathcal{Z}_a(z, v) \neq \emptyset \wedge |\mathcal{W}_a(z, v)| > 1)}_{(II)} \tag{B.8}$$

$$\tag{B.9}$$

The term (I) from (B.8) is absurd w.r.t. (2.32) and (2.34). The term (II) from (B.8) contradicts the assumption and concludes the proof. ∎

Proof of Lemma 4.1

The proof obviously follows from the fact that $\mathcal{Z}_s(0 \cdots K_e) = \emptyset$ is equivalent with $\lambda_s(z_s', z_s) = 0$, $\forall (z_s', z_s) \in \mathcal{Z}_p^2$ w.r.t. (4.24) for all types of specifications $\mathcal{S} \models Z_s$, $\mathcal{S} \models z_F$ and $\mathcal{S} \models W_s$ considered above. The characteristic function L_s of the specification automaton $\mathcal{N}_s$ then vanishes for every transition (z_p', w_p, z_p, v_p) of $\mathcal{N}_p$ because of (4.17). ∎

Proof of Theorem 4.1

The fact that $\mathcal{S} \models Z_s$ must first be feasible in order to be safely feasible is obvious. Therefore, under the assumption that basic feasibility of Z_s is given, only the equivalence to the safety condition needs to be proved. The following shows that it is impossible for $\mathcal{N}_p$ to deviate from Z_s despite its nondeterminism iff (4.28) holds.

($\Longrightarrow$) The negation of (4.28) is

$$\bigvee_{k=0}^{k_e-1} \bigvee_{z_p'}^{\mathcal{Z}_p \backslash z_{sk}'} \bigvee_{w_p}^{\mathcal{W}_p} \bigvee_{v_s}^{\mathcal{V}_{ap}(z_{sk}', z_{sk})} L_p(z_p', w_p, z_{sk}, v_s) = 1 \tag{B.10}$$

$$\Leftrightarrow \quad \exists k \in [0, k_e - 1], v_s \in \mathcal{V}_{ap}(z_{sk}', z_{sk}), z_p' \in \mathcal{Z}_p \backslash z_{sk}',$$

$$\text{and } w_p \in \mathcal{W}_p : L_p(z_p', w_p, z_{sk}, v_s) = 1 \tag{B.11}$$

Since Z_s is assumed to be basically feasible,

$$\exists w_p \in \mathcal{W}_p : L_p(z_{sk}', w_p, z_{sk}, v_s) = 1 \tag{B.12}$$

also hold. The fact that (B.11) and (B.12) simultaneously hold although $z_p' \in \mathcal{Z}_p \backslash z_{sk}' \Rightarrow z_p' \neq z_{sk}'$ reflects a deviation of the plant from $Z_s(k)$ to z_p' instead of $z_{sk}' = Z_s(k+1)$.

($\Longleftarrow$) If Z_s is not safely feasible then

$$\exists (w_{p1}, w_{p2}) \in \mathcal{W}_p^2, z_\sharp' \in \mathcal{Z}_p : L_p(z_{sk}', w_{p1}, z_{sk}, v_s) = 1 \text{ and } L_p(z_\sharp', w_{p2}, z_{sk}, v_s) = 1. \tag{B.13}$$

The last expression of (B.13) can be written as

$$\bigvee_{v_s}^{\mathcal{V}_{ap}(z_{sk}', z_{sk})} L_p(z_\sharp', w_{p2}, z_{sk}, v_s) = 1$$

which contradicts (4.28) and concludes the proof. ∎

Proof of Theorem 4.2

Recall that $\mathcal{Z}_s(0 \cdots K_e)$ is the state sequences set which is in line with $\mathcal{S} \models z_F$. The conditions to be fulfilled by a single state sequence $Z_s \in \mathcal{Z}_s(0 \cdots K_e)$ have already been proven in Theorem 4.1. The goal here is to show that it is necessary and sufficient for *all* state sequences of $\mathcal{Z}_s(0 \cdots K_e)$ to be safely feasible w.r.t. Theorem 4.1 in order for the corresponding specification $\mathcal{S} \models z_F$ to be safely feasible.

($\Longrightarrow$) First, $\mathcal{Z}_s(0 \cdots K_e)$ is derived from $\mathcal{S} \models z_F$ with (4.20). Since Definition 4.7 does not tolerate *any* violation of $\mathcal{S} \models z_F$ during its fulfillment in $\mathcal{N}_p$, it obviously concerns *every* state sequence $Z_s \in \mathcal{Z}_s(0 \cdots \boldsymbol{K_e})$ which is in line with $\mathcal{S} \models z_F$. Thus, the following holds: $\forall Z_s \in \mathcal{Z}_s(0 \cdots \boldsymbol{K_e})$, Z_s is safely feasible in $\mathcal{N}_p$ w.r.t. Theorem 4.1.

($\Longleftarrow$) Assume that $\exists Z_s \in \mathcal{Z}_s(0 \cdots \boldsymbol{K_e})$ which is not safely feasible w.r.t. Theorem 4.1. Then, (B.10)-(B.12) hold. As mentioned in the proof of Theorem 4.1, Eq.(B.11) and (B.12) reflect a deviation from the specification. This is a case of violation excluded by Definition 4.7. Hence, the corresponding specification $\mathcal{S} \models z_F$ is not safely feasible. ■

Proof of Theorem 4.3

The proof stems from [1]. Suppose that $\mathcal{N}_s$ describes a feasible specification w.r.t. Lemma 4.1. The generation of w_c is deterministic iff $\exists! w_c \in \mathcal{W}_c$ so that $L_c(*, w_c, z_c, v_c) = 1$ $\forall(z_c, v_c) \in \mathcal{Z}_c \times \mathcal{V}_c$. This is equivalent with

$$|\mathcal{W}_{ac}(z_c, v_c)| = 1. \tag{B.14}$$

Derive $\mathcal{V}_{as}(\cdot) = \mathcal{W}_{ac}(\cdot)$, $z_s = z_c$, and $w_s = v_c$, then apply it on (B.14) to obtain that $|\mathcal{V}_{as}(z_s, w_s)| = 1$ holds $\forall(z_s, w_s) \in \mathcal{Z}_s \times \mathcal{W}_s$, which is (4.47).

Equation (4.47) means that every output w_s must be generated by exactly one active input v_{as} from state z_s. It does not ask for z_s to have exactly one next state for each v_{as} i.e for the next state transition of $\mathcal{N}_s$ to be deterministic. Nor does it ask for every output w_s to be unique for each v_{as} from z_s i.e for the output generation of $\mathcal{N}_s$ to be deterministic. Hence, Theorem 4.3 shows that it is possible to deterministically control a nondeterministic specification automaton $\mathcal{N}_s$ for which (4.47) holds. ■

Proof of Theorem 4.4

($\Longrightarrow$) If (4.47) does not hold, then $\mathcal{N}_c$ is not W-deterministic. If $\mathcal{S}$ is not safely feasible w.r.t. Theorem 4.1, Corollary 4.1, and Theorem 4.2, there is no control law that can enforce $\mathcal{S}$ in $\mathcal{N}_p$; consequently there is no controller $\mathcal{N}_c$. In both cases $\mathcal{N}_p$ is not controllable w.r.t. $\mathcal{S}$ according to Definition 4.8.

($\Longleftarrow$) According to Definition 4.8, if $\mathcal{N}_p$ is not controllable w.r.t. $\mathcal{S}$ then $\mathcal{N}_c$ is not W-deterministic or the control loop is not weakly well-posed. $\mathcal{N}_c$ being not W-deterministic means that (4.47) does not hold. Theorem 4.3 then implies that $\mathcal{S}$ is not safely feasible.

The control loop not being weakly well-posed means that it is either well-posed or ill-posed. Without loss of generality, well-posedness is not possible without weakly well-posedness which is a relaxed property. The case of the ill-posedness described through (4.11) contradicts (4.47) because, e.g., $|\hat{\mathcal{W}}_c| = 0 \Leftrightarrow \exists (z_s, w_s) \in \mathcal{Z}_s \times \mathcal{W}_s : |\hat{\mathcal{V}}_{as}(z_s, w_s)| = 0$ by means of (4.3) and (4.17). This shows that (4.47) also does not hold in this case. Since the right side of the negated ($\Longleftarrow$) predicate to be proved is "$\mathcal{S}$ not safely feasible or (4.47) does not hold", the fact that the latter returns TRUE concludes the proof. ∎

C. MATLAB/Simulink programs: IDEFICS

This appendix gives an overview on the programs used in this thesis to investigate fault-tolerant control of nondeterministic I/O automata with computer-based means including the computation and analysis of complex automata and the visualization of selected behaviors. IDEFICS is the resulting toolbox and stands for the Implementation of Discrete-Event Fault tolerance In Control Systems. The programs have been developed and tested on MATLAB/Simulink R2012a version 7.14.0.739. The handbook [16] describes both MATLAB and Simulink programs. In the sequel, only MATLAB programs are presented. However, a complete version of IDEFICS version 1.0 is available for download under [11].

Section C.1 presents function used to create I/O automata in a monolithic and a component-oriented way. Functions required for control design purposes are given in Section C.2. Section C.3 contains functions where the operators, properties and analysis tools of Chapter 2 are implemented. Functions which can not be directly classified into modeling, control design or analysis but for specific purposes such encoding signals or computation observation are listed in Section C.4.

These programs were initially developed in [12] and continuously improved while working on this thesis. The code of some functions was revisited and extended in [132]. The supervised theses [17–24] and [25] contributed to further extensions of existing functions with additional features and the development of new ones.

C.1. Modeling of discrete-event systems

List of programs		
Function	Description	Page
buildnan	Creates a network of extended automata.	212
	...	

buildnan

Purpose Creates an automata network after the coupling matrix has been built with genkop.m and the automata have been generated with genat.m.

Syntax
```
ATNetwork = buildnan(K, AT1, ..., ATN, 'Name', 'NetName')
```

Input arguments

- K: Coupling matrix built with genkop.m.

- AT1-ATN are I/O automata to be built with genat.m.

Optional arguments

- Option: 'Name' (enter this if you want to name your network).

- Optionvalue: 'NetName' (enter the networks name as a string).

Output argument

- ATNetwork as a structure.

`completeAT`

Purpose Checks if an automaton is completely defined and completes it on demand.

Syntax `complete = completeAT(AT,mode)`

Input arguments

- AT: the automaton to check. This automaton must have been generated with the function genat().

Optional arguments

- mode: tells if the automaton should only be checked or also completed.
    - 0: check only the completeness (default).
    - 1: complete the automaton if it is incomplete.

Output arguments

- complete: will be either a boolean value or the completed automaton depending on the mode parameter.
    - 0: the automaton is incomplete (mode must be set to 0).
    - 1: the automaton is complete (mode must be set to 0).
    - The completed automaton (mode must be set to 1).

composeATN

Purpose Builds a composed I/O automaton from an automata network.

Syntax `ComposedAutomaton = composeATN(ATN,Option)`

Input argument

- ATN: is the automata network.

Optional arguments

- 'strict' for strictly well-posed automata networks only. The composition is aborted otherwise.
- 'weak' for weakly well-posed automata networks only. The composition is aborted otherwise.
- 'all' ignore the well-posedness property and execute the composition (default).

Output argument

- ComposedAutomaton as the composed I/O automaton.

computeAT

Purpose Simulates the behavior of an I/O automaton.

Syntax result = computeAT(AT,OPTION)

Input argument

- AT: is an I/O automaton without interconnection signals.

Optional arguments

- 'Name','SimulationName' is the name of the simulation to be given as a string.

- 'Z0', Z0 is the initial state of the automaton to be given as an integer Z0. If AT is a composed automaton, it is possible to give a row vector containing the initial states of each subautomaton. The order of the initial states in Z0 must be consistent with the order of automata saved in AT.attribute.consistsof.

- 'V',VSeq is the input sequence VSeq with the form:

$$\text{decoded :VSeq} = \begin{bmatrix} V_1(0) & V_1(1) & V_1(2) & \cdots \\ V_2(0) & V_2(1) & V_2(2) & \cdots \\ \vdots & \vdots & \vdots & \vdots \end{bmatrix}$$

$$\text{encoded :VSeq} = \begin{bmatrix} V(0) & V(1) & V(2) & \cdots \end{bmatrix}$$

- 'W', Wseq is the output sequence Wseq with the same form as Vfolge.

- Following combinations of Z,V and W are valid:

  - Z0, V for Simulation.

  - V und W for diagnosis.

  - Z0, V and W for diagnosis.

Output arguments

- result: contains the simulation results with the following fields:
    - trajectory: contains all possible trajectories.
    - mode: contains all parameters for plottrajectory.m.
    - ATName: the name of the automaton.
    - Name: the name of the simulation(only when existing).
    - consistent: contains boolean entries for each model which is consistent with the measurement (1=yes,0=no).
    - v: the input sequence used for simulation.
    - w: the output sequence used for simulation.
    - symbolset: the legend and description of states, input and output symbols.

decodezvw

Purpose Decodes a symbol of a composed I/O automaton.

Syntax decodedZVW = decodezvw(Content, zvw, AT)

Input arguments

- Content: symbol to decode.
- zvw: string representing the symbol to decode with 'z' for a state, 'v' for an input and 'w' for an output.
- AT: composed automaton as a structure resulting from composeATN.m.

Output argument

- decodedZVW: decoded state, input or output.

doub2int

Purpose Converts a matrix of type double into a matrix of type integer of the adequate integer-type w.r.t. the maximal and minimal values. Negative values are assumed not to be smaller than -128.

Syntax `intMatrix = doub2int(doubMatrix)`

Input argument

- doubMatrix: Matrix with entries of the data type double.

Output argument

- intMatrix: Matrix with entries of the data type integer.

genat

Purpose This function creates classic and extended I/O automata. The calls 1) and 2) return tables of the automaton. The call 3) returns structures containing tables and other information for component oriented modeling.

Syntax

Call 1): `automaton = genat(F,G,H,z0)`

Call 2): `automaton = genat(Nz, Nv, Nw, Ns, Nr, z0)`

An interactive program is started with a command prompt. The user can enter desired tuple. Only consistent tuples are accepted.

Call 3): `automaton = genat(table, Dim, z0, 'Name', 'NameOfTheAT')`

Input arguments

Call 1):

- F, G and H are tables with 4 columnns of the following structure: F -> [r z v s], G -> [z' z v s], H -> [w z v s]

- z0 is the initial state.

Call 2):

- Nz, Nv, Nw, Ns, Nr are vectors representing the symbol sets.

- z0 is the initial state.

Call 3):

- table: automaton table with the following columns [Z'cols Wcols Rcols Zcols Vcols Scols]

- Dim: the dimension vector with the columns [dim_z dim_w dim_r dim_v dim_s]

- z0 is the initial state. If z0 is a vector, it is encoded into a scalar by the program.

- 'Name' is an optional parameter where 'NameOfTheAT' is the assigned value to give the automaton a name.

Output arguments

Call 1) and 2): automaton : the automaton table with the columns
[z' w r z v s]

Call 3): The automaton is a structure with the following fields:

- .attribute.

    - .symbolset: sets of all epsilon symbols.

    - .type: single (default) or composed automaton.

    - .dim (if the table is encoded): dimension vector once encoded.

    - .rdim (if the table is encoded): initial dimension before encoding.

    - .det: 1 if deterministic, 0 otherwise.

    - .complete: 1 if completely defined, 0 otherwise.

    - .name: given name.

- .symbolset: sets of all symbols.

    - .z.

    - .w.

    - .r.

    - .v.

    - .s.

- .table: table with the defined transitions.

genkop

Purpose Builds the coupling matrix of an I/O automata network.

Syntax K = genkop(rvector, n_rPorts, n_sPorts)

Input arguments

- rvector: contains the index of r-ports which are connected to the corresponding s-ports according to their position in the vector. If the i-th entry of the rvector is 0, then the i-th s-port is not connected (this happens e.g. serial connections).

- n_rPorts: number of r-ports or the number of columns of the coupling matrix.

- n_sPorts: number of s-ports or the number of rows of the coupling matrix.

Output argument

- K: coupling matrix of an I/O automata network.

normAT

Purpose Normalizes the table of an I/O automaton. To normalize means to have symbols ranging from 1 to |Zset| for states, |Vset| for inputs and |Wset| for outputs.

Syntax [Norm_AT Legend] = normAT(AT,zvw)

Input argument

- AT: I/O automaton as a matrix with the columns [z' w z v].

Optional arguments

- zvw: specifies the symbols to be normalized as follows:
  - 'zvw' > states, inputs and outputs (default).
  - 'z' > states only.
  - 'w' > outputs only.
  - 'zv' > states and inputs only.
  - ... > any combination in any order of z,v and w is allowed.

Output argument

- Norm_AT: normalized I/O automaton.

- Legend: Structure containing tables where the left column consists of the original symbols and the right column the new symbols as follows:
  - Legend.z > Legend of normalized states
  - Legend.v > Legend of normalized inputs
  - Legend.w > Legend of normalized outputs

  The legend of symbols which are not formalized is empty.

simAT

Purpose Simulates the behavior of an I/O automaton from an initial state with an input sequence.

Syntax [Zseq Wseq] = simAT(AT, z0, Vseq)

Input arguments

- AT: I/O automaton as a matrix with the columns [z' w z v].

- z0: initial state.

- Vseq: input sequence of the I/O automaton as a vector.

Output arguments

- Zseq: states sequence in a matrix where each line corresponds to a possible state sequence.

- Wseq: output sequence in a cell-array where each cell corresponds to a state sequence.

simLpf

Purpose Computes the model of a faulty plant as an I/O automaton.

Syntax `ATf = simLpf(AT,Errz,Errv,Errw)`

Input arguments

- AT: I/O automaton as a matrix with the columns [z' w z v] representing the fault free system.

- Errv: relation between the correct and the wrong inputs in a table with two columns [vf vn]. vn is the column of nominal inputs and vf is the column of wrong inputs to consider when the input signal vn is received for the same line in Errv. Failures are symbolized by vf = -5 and lead to self-loops (z'=z).

- Errz: analog to Errv with [zf z v] tables.

- Errw: relation between the correct and the wrong outputs in a table with two columns [wf wn]. wn is the column of nominal outputs and wf is the column of wrong outputs. Failures are symbolized by wf = -5.

Output argument

- ATf: I/O automaton as a matrix with the columns [z' w z v] representing the faulty system.

Note. Since this is an exact implementation of Eq. (1.19) in [8], some complexity limitations might be encountered. If needed, pre-allocate a space of the size $|W| \times |V| \times (|Z|^2)$.

simulateATN

Purpose Simulates the behavior of an I/O automata network.

Syntax `result = simulteATN(ATN,Z0,V,mode)`

Input arguments

- ATN: is the I/O automata network.

- Z0: is the initial state of the automata network. It is row vector containing the initial state of each I/O automaton in the network. The order of the initial states in Z0 must be consistent with the order of automata saved in ATN.attribute.consistsof.

- V: is the input sequence with the form:

$$
V = \begin{bmatrix} V_1(k=0) & V_1(k=1) & V_1(k=2) & \cdots \\ V_2(k=0) & V_2(k=1) & & \cdots \\ V_3(k=0) & & \cdots & \end{bmatrix}
$$

Optional argument

- mode: is a parameter which permits to selects the well-posedness condition under which the on-line composition during the simulation should be checked. Possible values are:

    - 'strict' for strictly well-posed automata networks only. The composition is aborted otherwise.

    - 'weak' for weakly well-posed automata networks only. The composition is aborted otherwise.

    - 'all' ignore the well-posedness property and execute the composition (default).

Output argument

- result: saves the result of the simulation in a structure with the following fields:

    - .trajectory: States sequence.

    - .v: input sequence.

- .consistent: boolean value revealing if the execution the whole sequence was successful or not: 1-> Yes, 0-> No.

- .symbolset: set of all symbols.

- .mode: required for visualization purposes with plottrajectory.m.

- .ATName: name of the automata network.

- .Name: name of the simulation.

simWapf

Purpose Computes the eventually faulty outputs of an I/O automaton.

Syntax Wapf = simWapf(AT,Zknextset,Zkset,Errw)

Input arguments

- AT: I/O automaton as a matrix with the columns [z' w z v] representing the fault free system.

- Zknextset: cell array with possible next states of the I/O automaton in each cell. The cell Zknextseti contains the next states of the vector Zkset(i). This parameter can be computed with simG or simZapf. It must be a 1-column cell.

- Zkset: vector with the possible current states of the I/O automaton.

- Errw: relation between the correct and the wrong outputs in a table with two columns [wf wn]. wn is the column of nominal outputs and wf is the column of wrong outputs. Failures are symbolized by wf = -5.

Output argument

- Wapf: a cell array containing in Wapfi,j the set of faulty outputs states for each state zk = Zset(i) and each next state Zknextseti(j).

simZapf

Purpose Computes the eventually faulty next states of an I/O automaton.

Syntax `Zapf = simZapf(AT,Zsetk,vk,Errz,Errv)`

Input arguments

- AT: I/O automaton as a matrix with the columns [z' w z v] representing the fault free system.

- Errv: relation between the correct and the wrong inputs in a table with two columns [vf vn]. vn is the column of nominal inputs and vf is the column of wrong inputs to consider when the input signal vn is received for the same line in Errv. Failures are symbolized by vf = -5 and lead to self-loops (z'=z).

- Errz: analog to Errv with [zf z v] tables. The failure state is symbolized by zf = -9.

Output argument

- Zapf: a cell array containing in Zapfi the set of faulty next states for each state zk = Zset(i). If a state-input combination is not specified in Errz, but results from the evaluation of Errv(vk) with zk, then the current next state is returned by `simG(zk,Errv(vk),AT)`.

C.2. Discrete-event control design

List of programs		
Function	Description	Page
buildcontroller	Computes the supercontroller Nc.	226
buildNs	Computes the subautomaton.	227
decloop	Simulates the behavior of a discrete-event control loop.	228
	...	

List of programs (Part II)		
Function	Description	Page
decomposeNc	Decomposes a supercontroller Nc.	229
reconf	Computes a new controller, a new state sequence and a new counter value based on the diagnosis result.	229

`buildcontroller`

Purpose Computes the supercontroller Nc for a plant Np and a specification S of the type Stype.

Syntax Nc = buildcontroller(Np,S,Stype)

Input arguments

- Np: I/O automaton modeling the plant to control. It has the columns form [z' w z v] and the initial state $z0p = Np(1,3)$.

- S: specification to be achieved by the plant in the closed loop with the controller.

- Stype: Selected type of the specification out the following:

  - 'zF': The plant should reach the final state zF from the initial state z0p. S must be a scalar.

  - 'Zs': The plant should follow a given state sequence $Zs(0...ke) = z0p \times Zs(1...ke)$. S must be a vector of the length ke+1 containing the state sequence $Zs(0...ke)$.

  - 'Ws': The plant should generate a given output sequence $Ws(0...ke)$. S must be a vector of the length ke+1 containing the output sequence $Ws(0...ke)$.

  - 'Zill': The plant should avoid the illegal states in Zill. S must be a vector containing those forbidden states.

  - 'Will': The plant should avoid generating the illegal outputs in Will. S must be a vector containing those forbidden outputs.

- 'Till': The plant should avoid executing the illegal transitions in Till. S must be a matrix containing those forbidden transitions as rows with the following columns [z' w z v].

Output argument

- Nc: Supercontroller as a cellarray containing a list of control laws which can be used to achieve the specification.

buildNs

Purpose Computes the subautomaton of the plant model Np which is consistent with the given specification S.

Syntax Ns = buildNs(Np,S,Stype)

Input arguments

- Np: I/O automaton modeling the plant to control. It has the columns form [z' w z v] and the initial state $z0p = Np(1,3)$.

- S: specification to be achieved by the plant in the closed loop with the controller.

- Stype: Selected type of the specification out the following:

    - 'zF': The plant should reach the final state zF from the initial state z0p. S must be a scalar.

    - 'Zs': The plant should follow a given state sequence $Zs(0...ke) = z0p \times Zs(1...ke)$. S must be a vector of the length ke+1 containing the state sequence $Zs(0...ke)$.

    - 'Ws': The plant should generate a given output sequence $Ws(0...ke)$. S must be a vector of the length ke+1 containing the output sequence $Ws(0...ke)$.

    - 'Zill': The plant should avoid the illegal states in Zill. S must be a vector containing those forbidden states.

- – 'Will': The plant should avoid generating the illegal outputs in Will. S must be a vector containing those forbidden outputs.
- – 'Till': The plant should avoid executing the illegal transitions in Till. S must be a matrix containing those forbidden transitions as rows with the following columns [z' w z v].

Output argument

- Ns: subautomaton of Np which is consitent with the specification S.

`decloop`

Purpose Simulates the behavior of a discrete-event control loop.

Syntax [Zc Wc Zp Wp] = decloop(Ac, z0c, Np, z0p, ke)

Input arguments

- Ac: control law as an I/O automaton.

- z0c: initial state of the controller.

- Np: model of the plant.

- z0p: initial state of the plant.

- ke: maximal simulation step.

Output arguments

- Zc: State sequence of the controller.

- Wc: Output sequence of the controller.

- Zp: State sequence of the plant.

- Wp: Output sequence of the plant.

decomposeNc

Purpose Decomposes a supercontroller Nc into a set of control laws Ac ^ (i)'s for a given state sequence.

Syntax `[Acis AcPerZs] = decomposeNc(Nc,Zshat, nu)`

Input arguments

- Nc: supercontroller containing all control laws.

- Zshat: List of state sequence as a cellarray where $Zs\_i = Zshati$.

- nu: number of control laws.

Output arguments

- Acis: all computed control laws.

- AcPerZs: List of index i of Acis computed per state sequence Zs of Zshat.

reconf

Purpose Computes a new controller, a new state sequence and a new counter value based on the diagnosis result. automaton.

Syntax `[Acr Zsr kc] = reconf(Ac, Zs,diag)`

Input arguments

- Ac: Control law of the nominal plant. It has the columns form [z' w z v].

- Zs: set trajectory of the controller in the nominal case.

- Zkset: vector with the possible current states of the I/O automaton.

- diag: the diagnosis report as a structure with the following fields

- .Npf: Model of the faulty plant with the columns [z' w z v].

- .fcode: A number between 0 and 7 with the following meaning: 0=faultfree, 1=actuator fault, 2=internal fault, 3=sensor fault, 4=actuator failure, 5=internal failure, 6=sensor failure, 7=unknown fault.

- .faultysig: cell array with the faulty signals

  faultysig{1}=vpf (the faulty input)

  faultysig{2}=zpf (the faulty state)

  faultysig{3}=wpf (the faulty output)

- .nominalsig: cell array with the nominal i.e. expected signals

  nominalsig{1}=vpn (the expected input)

  nominalsig{2}=zpn (the expected state)

  nominalsig{3}=wpn (the expected output)

- .errorfunc: cell array with the error functions

  errorfunc{1}=Errv (the input error relation)

  errorfunc{2}=Errz (the state error relation)

  errorfunc{3}=Errw (the output error relation)

- .ksf: value of the internal counter of the controller at the moment of the fault.

- .ki: identification step.

Output arguments

- Acr: reconfigured controller.

- Zsr: reconfigured state sequence.

- kc: new value of the internal value of the controller.

C.3. Analysis operators

List of programs		
Function	Description	Page
activeInput	Returns the set of active inputs events.	232
activeNextStates	Returns the set of active next states.	233
activeOutput	Returns the set of active outputs events.	234
buildAv	Builds the symbolic input adjacency matrix.	234
buildAw	Builds the symbolic output adjacency matrix.	235
buildAz	Builds the symbolic state adjacency matrix.	235
comparetrajectories	Compares simulation results.	236
diagnoser	Detects and identify faulty transitions and faulty outputs.	236
getAG	Computes the state transition adjacency or reachability matrix.	237
getfg	Computes the degree of freedom.	238
getrddeg	Computes the redundancy degree.	238
getZposs	Generates the possible states sequences.	239
incidentInputs	Computes the set of incident inputs.	239
isblocking	Checks if a discrete-event control loop will block.	240
isdet	Checks if an extended automaton is deterministic.	241
isdeter	Checks if an I/O automaton is G-deterministic and H-deterministic.	241
isGdet	Checks if an I/O automaton is G-deterministic.	242
isHdet	Checks if an I/O automaton is H-deterministic.	243
ishomomorph	Checks the homomorphism between two automata.	243
mexGHz0	Simulates the behavior of an I/O automaton.	245
	...	

List of programs (Part II)		
Function	Description	Page
outdeg	Computes the output degree.	245
plottrajectory	Helps to visualize trajectories of I/O automata.	246
reachable	Deletes the nonreachable states.	246
simG	Simulates the state transition function.	247
simH	Simulates the output generation function.	247
simL	Implements the boolean characteristic function.	248
strongconsets	Computes the set of strongly connected states.	248

activeInput

Purpose Returns the set of active inputs events for a state or state transition in an I/O automaton.

Syntax Vactive = activeInput(zvwStr,zvwVal,AT)

Input arguments

- zvwStr: specifies a signal combination out of the following:
    - 'z': single state.
    - 'nzz': nextstate-state combination.
    - 'zw': state-output combination.
    - 'nzzw': nextstate-state-output combination.
- zvwVal: values of the symbols according to zvwStr:
    - z: scalar value representing the state.
    - [z' z]: vector with the nextstate and the initial state as scalars.
    - [z w]: vector with the state and the output as scalars.
    - [z',z,w]: vector with the nextstate, the initial state and the input and the output as scalars.

- AT: I/O automaton to consider.

Output argument

- Vactive: set of active inputs for the selected symbols combination.

activeNextStates

Purpose Returns the set of active next states from a given state in an I/O automaton.

Syntax Zactive = activeNextStates(zvwStr,zvwVal,AT)

Input arguments

- zvwStr: specifies a signal combination out of the following:

    - 'z': single state.
    - 'zv': state-input combination.
    - 'zw': state-output combination.
    - 'zvw': state-input-output combination.

- zvwVal: values of the symbols according to zvwStr:

    - z: scalar value representing the state.
    - [z v]: vector with the state and the input as scalars.
    - [z w]: vector with the state and the output as scalars.
    - [z,v,w]: vector with the state, the input and the output as scalars.

- AT: I/O automaton to consider.

Output argument

- Zactive: set of active next states for the selected symbols combination.

activeOutput

Purpose Returns the set of active outputs events for a state or state transition in an I/O automaton.

Syntax Wactive = activeOutput(zvwStr,zvwVal,AT)

Input arguments

- zvwStr: specifies a signal combination out of the following:
    - 'z': single state.
    - 'nzz': nextstate-state combination.
    - 'zv': state-input combination.
    - 'nzzv': nextstate-state-input combination.
- zvwVal: values of the symbols according to zvwStr:
    - z: scalar value representing the state.
    - [z' z]: vector with the nextstate and the initial state as scalars.
    - [z v]: vector with the state and the input as scalars.
    - [z',z,v]: vector with the nextstate, the initial state and the input as scalars.
- AT: I/O automaton to consider.

Output argument

- Wactive: set of active outputs for the selected symbols combination.

buildAv

Purpose Builds the symbolic input adjacency matrix Av for a given I/O automaton AT.

Syntax `Av = buildAv(AT)`

Input argument

- AT: I/O automaton with the following columns form [z' w z v] where the states are assumed to be labeled from 1 to |Z|. Eventuallly normalize the automaton table with normAT().

Output argument

- Av: symbolic input adjacency matrix of AT.

buildAw

Purpose Builds the symbolic output adjacency matrix Aw for a given I/O automaton AT.

Syntax `Aw = buildAw(AT)`

Input argument

- AT: I/O automaton with the following columns form [z' w z v] where the states are assumed to be labeled from 1 to |Z|. Eventuallly normalize the automaton table with normAT().

Output argument

- Aw: symbolic output adjacency matrix of AT.

buildAz

Purpose Builds the symbolic state adjacency matrix Az for a given I/O automaton AT.

Syntax `Az = buildAz(AT)`

Input argument

- AT: I/O automaton with the following columns form [z' w z v] where the states are assumed to be labeled from 1 to |Z|. Eventuallly normalize the automaton table with normAT().

Output argument

- Az: symbolic state adjacency matrix of AT.

comparetrajectories

Purpose Compares simulation results of the same composed I/O automaton.

Syntax `match = comparetrajectories(result_1,result_2,...)`

Input argument

- result_n the n-th simulation result.

Output argument

- match tells if the trajectories match or not.

diagnoser

Purpose Detects and identify faulty transitions and faulty outputs in an automaton based on the I/O measurements and the observed state sequence.

Syntax

```
[isFaulty fStep fIdent] = diagnoser(VsSeq, ZsSeq, WsSeq,
AT)
```

Input arguments

- VsSeq: Input sequence.

- ZsSeq: Observed state sequence.

- WsSeq: Output sequence.

- AT: I/O automaton to consider.

Output arguments

- isFaulty: boolean value of the fault detection with 1 = faulty and 0 = no fault.

- fStep: step where the faulty transition was detected.

- fIdent: Identification of the fault:

    - [1 zf]: zf is the faulty state.

    - [2 wf]: wf is the faulty output

getAG

Purpose Computes the state transition adjacency or reachability matrix with variable horizon for any possible input.

Syntax `AG = getAG(AT, horizon)`

Input arguments

- AT: I/O automaton as a matrix with the columns [z' w z v].

- horizon: number of steps to compute the reachability for.

Output argument

- AG: reachability matrix for the given number of steps in AT.

getfg

Purpose Computes the degree of freedom of the specification Stype zF.

Syntax deg = getfg([z0 zF],AT)

Input arguments

- [z0 zF]: vector with the specified initial state z0 and the final state zF.

- AT: I/O automaton as a matrix with the columns [z' w z v].

Output argument

- deg: degree of freedom.

getrddeg

Purpose Computes the redundancy degree between two states of a supercontroller.

Syntax [rdDeg Zposs Wposs Vposs] = getrddeg(Nc, z0, zc)

Input arguments

- Nc: supercontroller as a matrix with the columns [z' w z v].

- z0: initial state.

- zc: target state in Nc.

Output arguments

- rdDeg: redundancy degree.

- Zposs: possible state sequences.

- Wposs: possible output sequences.

- Vposs: possible input sequences.

getZposs

Purpose Generates the possible states sequences which can be performed by an automaton AT starting at z0 and ending at zF.

Syntax `Zposs = getZposs(z0, zF, AT)`

Input arguments

- z0: initial state.

- zF: final state zF.

- AT: I/O automaton as a matrix with the columns [z' w z v].

Output argument

- Zposs: a cell-array containing all possible states sequences.

incidentInputs

Purpose Computes the set of incident inputs to a given state of an I/O automaton.

Syntax `VincidentZ = incidentInputs(z,AT)`

Input arguments

- z: state to consider.

- AT: I/O automaton as a matrix with the columns [z' w z v].

Output argument

- VincidentZ: set of incident inputs to state z in AT.

isblocking

Purpose Checks if a discrete-event control loop will block within a given finite number of steps.

Syntax
```
[blocking transitions step] = isblocking(Ac, z0c, Np, z0p,
ke)
```

Input arguments

- Ac: control law as a matrix with the columns [z' w z v].

- z0c: initial state of the controller.

- Np: model of the plant as a matrix with the columns [z' w z v].

- z0p: initial state of the plant.

- ke: maximal simulation step.

- zF: final state of the plant to consider.

Output arguments

- blocking: 1='blocking' and 0='nonblocking'.

- transitions: Matrix of the form [vc zc wc zp].

- step: blocking step of the control loop.

isdet

Purpose Checks if an automaton is deterministic.

Syntax [isdet transitions] = isdet(AT)

Input argument

- AT is an extended I/O automaton obtained with the function genat or composeATN.

Output arguments

- isdet is a boolean value
  - 1: the automaton is deterministic
  - 0: the automaton is nondeterministic
- transistions contains a list of the nondeterministic transitions.

isdeter

Purpose Checks if an I/O automaton is G-deterministic and H-deterministic. An I/O automaton is G-deterministic if its state transition function G is deterministic. An I/O automaton is H-deterministic if its output generation function H is deterministic.

Syntax [DetFlag nDetGTrans nDetHTrans] = isdeter(AT)

Input argument

- AT: I/O automaton as a matrix with the columns [z' w z v].

Output arguments

- DetFlag: 1=AT is deterministic and 0=AT is nondeterministic.

- nDetGTrans: the list of G-nondeterministic transitions i.e those where there exist more than one next state for a given state-input combination. Trans = AT(nDetGTrans1(1),:) is a transition for which z=Trans(3) and v=Trans(4) lead to more than one next state z'.

- nDetHTrans: the list of H-nondeterministic transitions i.e those where there exist more than one output for a given state-input combination. Trans = AT(nDetHTrans1(1),:) is a transition for which z=Trans(3) and v=Trans(4) lead to more than one output w.

isGdet

Purpose Checks if an I/O automaton is G-deterministic. An I/O automaton is G-deterministic if its state transition function G is deterministic.

Syntax [GDet Gtrans] = isGdet(AT)

Input argument

- AT: I/O automaton as a matrix with the columns [z' w z v].

Output arguments

- GDet: 1=AT is G-deterministic and 0=AT is G-nondeterministic.

- Gtrans: list of row number in AT which are G-nondeterministic i.e those where there exist more than one next state for a given state-input combination. Trans = AT(Gtrans1(1),:) is a transition for which z=Trans(3) and v=Trans(4) lead to more than one next state z'.

isHdet

Purpose Checks if an I/O automaton is H-deterministic. An I/O automaton is H-deterministic if its output generation function H is deterministic.

Syntax [HDet Htrans] = isHdet(AT)

Input argument

- AT: I/O automaton as a matrix with the columns [z' w z v].

Output arguments

- HDet: 1=AT is H-deterministic and 0=AT is H-nondeterministic.

- Htrans: list of row number in AT which are H-nondeterministic i.e those where there exist more than one next state for a given state-input combination. Trans = AT(Htrans1(1),:) is a transition for which z=Trans(3) and v=Trans(4) lead to more than one output w.

ishomomorph

Purpose Checks if an I/O automata AT2 is a homomorph image of ab I/O automaton AT1.

Syntax hom = ishomomorph(AT1,AT2)

Input arguments

- AT1: left-hand side of the homomorphous map.

- AT2: right-hand side of the homomorphous map.

Output argument

- hom: 1=yes and 0=no.

mexGHz0

Purpose Simulates the behavior of an I/O automaton.

Syntax [zSeq wSeq typSeq] = mexGHz0(AT, z0, vSeq)

Input arguments

- AT: I/O automaton with columns of the form | z' | w | z | v |.

- z0: Initial state.

- vSeq: input sequence.

Output arguments

- zSeq: states sequence.

- wSeq: output sequence.

- typSeq: (optional) a sequences of flags which can be used to interpret the transition saved in zSeq and wSeq.

    - 0 : invalid transition at the corresponding step although.

    - 1 : valid transition.

    - 2 : One or several symbols do not exist in 'AT'. A warning is sent out.

outdeg

Purpose Computes the output degree of a state or state combination in an I/O automaton.

Syntax od = outdeg(znzVektor, AT)

Input arguments

- znzVektor: single state z or couple of states [z' z] where z' is the next state to consider.

- AT: I/O automaton as a matrix with the columns [z' w z v]. Make sure that there are no double entries in AT, otherwise wrong values would be returned.

Output argument

- od: computed output degree.

plottrajectory

Purpose Helps to visualize trajectories of I/O automata.

Syntax `plottrajectory(result,faultmodel)`

Input argument

- result: the simulation result to be visualized.

Optional argument

- faultmodel: the model of the fault from which the trajectory of the automaton should be plotted. If this parameter is missing, all trajectories are plotted in separated windows.

reachable

Purpose Deletes the nonreachable states of AT except the initial states of Z0.

Syntax `[reachableAT nonreachZ] = reachable(AT, Z0)`

Input arguments

- AT: I/O automaton as a matrix with the columns [z' w z v].
- Z0: set of initial states.

Output arguments

- reachableAT: I/O automaton containing only reachable states.

- nonreachZ: Set of nonreachable states (Z0 excluded).

simG

Purpose Simulates the state transition function G of an I/O automaton.

Syntax `zn = simG(z,v,AT)`

Input arguments

- z: state.

- v: input.

- AT: I/O automaton as a matrix with the columns [z' w z v].

Output argument

- zn: next state z'.

simH

Purpose Simulates the output generation function H of an I/O automaton.

Syntax `w = simH(z,v,AT)`

Input arguments

- z: state.

- v: input.

- AT: I/O automaton as a matrix with the columns [z' w z v].

Output argument

- w: output.

simL

Purpose Implements the boolean characteristic function of an I/O automaton.

Syntax `Lvalue = simL(AT,ZnextSeq, Wseq, Zseq, Vseq)`

Input arguments

- AT: I/O automaton as a matrix with the columns [z' w z v].

- ZnextSeq: List of next state sequences where e.g ZnextSeq(2,5) is a next state of the second sequence at step 5.

- Wseq: List of output sequences similarly built as ZnextSeq.

- Zseq: List of state sequences similarly built as ZnextSeq (must have one element more than the other sequences).

- Vseq: List of input sequences similarly built as ZnextSeq.

Output argument

- Lvalue: value of the characteristic function. Lvalue=1 means that the corresponding sequences are feasible in AT, otherwise Lvalue=0.

strongconsets

Purpose Computes the set of strongly connected states.

Syntax `Sets = strongconsets(AT)`

Input argument

- AT: I/O automaton as a matrix with the columns [z' w z v].

Output argument

- Sets: set of strongly connected states in a cell-array.

C.4. Specific features

List of programs		
Function	Description	Page
convertSequence	Encodes and decodes sequences.	250
convertStrSet	Makes conversion between a string cell-arrays and a string chain.	250
encodeSets	Encodes the given sets of symbol set structures.	251
encodeSetsClean	Encodes only existing tuples in the given sets of symbol set structures.	252
getvset	Gives the set of inputs of an I/O automaton.	253
getwset	Computes the set of outputs of an I/O automaton.	253
getzset	Computes the set of states of an I/O automaton.	253
karteprod	Implements the cartesian product operator for matrices.	254
karteprodint	Implements the cartesian product operator for integer matrices.	254
ncmultiply	Raises a symbolic matrix to a given power.	255
numtab2strtab	Converts a numerical table into a string table.	256
replaceDC	Replaces all 'don't care' symbols.	256
	...	

List of programs (Part II)		
Function	Description	Page
waitbarx	Shows the evolution of a computation.	257
zposs2adj	Generates the adjacency matrix.	257

convertSequence

Purpose Encodes and decodes a state, input or output sequence.

Syntax `convertedSequence = convertSequence(codingLaw, sequence)`

Input arguments

- coding: is the coding law of the automaton generated during its creation.

- sequence: the sequence to be encoded or decoded. Sequences to be encoded should have the following form e.g. for an input sequence VSeq: $V_{Seq} =$

$$\begin{bmatrix} V_1(0) & V_2(0) & V_3(0) & \cdots \\ V_1(1) & V_2(1) & V_3(1) & \cdots \\ V_1(2) & V_2(2) & V_3(2) & \cdots \\ \vdots & \vdots & \vdots & \vdots \end{bmatrix}$$

The program autonomously detects when to encode or to decode.

Output argument

- convertedSequence is the encoded or decoded sequence.

convertStrSet

Purpose Converts a multiple row string cell-array into a string chain and vice versa.

Syntax `strSet = convertStrSet(cellSet)`

Input arguments

- cellSet/strSet is the string or cell-array to convert with the following form:

$$- \text{ cellSet: } \begin{bmatrix} \text{String11} & \text{String12} & \cdots \\ \text{String21} & \cdots & \\ \vdots & \vdots & \end{bmatrix}$$

$$- \text{ strSet: } \begin{bmatrix} \text{'String11'} & \text{'String12'} \cdots \\ \text{'String21'} & \cdots & \\ \vdots & \vdots & \end{bmatrix}$$

Output arguments

- strSet/cellSet as in the input arguments.

Input parameter: strSet -> Output parameter: cellSet
Input parameter: cellSet -> Output parameter: strSet

encodeSets

Purpose Encodes the given sets of symbol set structures.

Syntax `set = encodeSets(sets)`

Input argument

- sets: contains the symbol set structure to be encoded and has the form:

$$\begin{bmatrix} |\text{V1}| & \begin{bmatrix} v_1 & v_2 & \cdots \end{bmatrix} & \begin{bmatrix} \text{'str1'} & \text{'str1'} & \cdots \end{bmatrix} & \text{'V1 Name'} \\ |\text{V2}| & \begin{bmatrix} v_1 & v_2 & \cdots \end{bmatrix} & \begin{bmatrix} \text{'str1'} & \text{'str1'} & \cdots \end{bmatrix} & \text{'V2 Name'} \\ \vdots & \vdots & \vdots & \vdots \end{bmatrix}$$

Output argument

- set   contains   the   symbol   set   structure   with   the   form:

$$\left[|V| \quad \begin{bmatrix} 1 & 2 & \cdots \end{bmatrix} \quad \begin{bmatrix} \text{'str1'} & \text{'str1'} & \cdots \end{bmatrix} \quad \text{coding} \right]$$

Coding is encoding table.

encodeSetsClean

Purpose Encodes the given sets of symbol set structures. In contrast to encodeSets.m, this function encodes only tuples which exist in the given table.

Syntax `set = encodeSetsClean(sets,table)`

Input arguments

- sets: contains the symbol set structure to be encoded and has the form:

$$\begin{bmatrix} |V1| & \begin{bmatrix} v_1 & v_2 & \cdots \end{bmatrix} & \begin{bmatrix} \text{'str1'} & \text{'str1'} & \cdots \end{bmatrix} & \text{'V1 Name'} \\ |V2| & \begin{bmatrix} v_1 & v_2 & \cdots \end{bmatrix} & \begin{bmatrix} \text{'str1'} & \text{'str1'} & \cdots \end{bmatrix} & \text{'V2 Name'} \\ \vdots & \vdots & \vdots & \vdots \end{bmatrix}$$

- table: Table with valid combinations.

$$\begin{bmatrix} v_1 & v_2 & v_3 & \cdots \end{bmatrix}$$

Output argument

- set   contains   the   symbol   set   structure   with   the   form:

$$\left[|V| \quad \begin{bmatrix} 1 & 2 & \cdots \end{bmatrix} \quad \begin{bmatrix} \text{'str1'} & \text{'str1'} & \cdots \end{bmatrix} \quad \text{coding} \right]$$

Coding is encoding table.

getvset

Purpose Gives the set of inputs of an I/O automaton.

Syntax vset = getvset(AT)

Input argument

- AT: I/O automaton as a matrix with the columns [z' w z v].

Output argument

- vset: set of inputs in AT.

getwset

Purpose Computes the set of outputs of an I/O automaton.

Syntax wset = getwset(AT)

Input argument

- AT: I/O automaton as a matrix with the columns [z' w z v].

Output argument

- wset: set of outputs in AT.

getzset

Purpose Computes the set of states of an I/O automaton.

Syntax `zset = getzset(AT)`

Input argument

- AT: I/O automaton as a matrix with the columns [z' w z v].

Output argument

- zset: set of states in AT.

karteprod

Purpose Implements the cartesian product operator for matrices.

Syntax
`ResultMatrix = karteprod(Matrix1, Matrix2, ..., MatrixN)`

Input argument

- MatrixI: matrice I to be multiplied via the cartesian operator. Vectors have to be entered column wise. Row vectors are not multiplied element-wise but block wise.

Output argument

- ResultMatrix: the cartesian product of all entered matrices.
  ResultMatrix = Matrix1 X Matrix2 X ... X MatrixN.

karteprodint

Purpose Implements the cartesian product operator for integer matrices. The difference with the function karteprod() is that karteprodint() produces an integer matrix instead of a double matrix as karteprod() does.

Syntax
```
ResultMatrix = karteprodint(Matrix1, Matrix2, ..., MatrixN)
```

Input argument

- MatrixI: matrice I to be multiplied via the cartesian operator. Vectors have to be entered column wise. Row vectors are not multiplied element-wise but block wise. This matrices can be converted from double type to integer type with the function doub2int().

Output argument

- ResultMatrix: the cartesian product of all entered matrices.
  ResultMatrix = Matrix1 X Matrix2 X ... X MatrixN.

ncmultiply

Purpose Implements the cartesian product operator for matrices. The algorithm executes a noncommutative matrix multiplication.

Syntax `MatrixOut = ncmultiply(MatrixIn,power)`

Input arguments

- MatrixIn: symbolic adjacency matrix to be raised to the given power.

- power: power to which the matrix should be raised.

Output argument

- MatrixOut: resulting symbolic adjacency matrix.

numtab2strtab

Purpose Converts a numerical table into a string table with labels for each column in the header of the string table.

Syntax
```
stringTab = numtab2strtab(numTab,HeaderString,NbrOfBlanks)
```

Input arguments

- numTab: the numerical table with N columns.

- HeaderString: cell array of N string containing the title of each column.

- NbrOfBlanks: it is optional. It is the number of blanks between the columns.

Output argument

- stringTab: the string table with titles at the top of each column.

replaceDC

Purpose Replaces all 'don't care' symbols and deletes double entries.

Syntax `newTable = replaceDC(Table,dim,dontcare,epsilon)`

Input arguments

- Table: the table of the extended I/O automaton with columns of the form:
$$\begin{bmatrix} z' & w & r_1 - r_m & z & v & s_1 - s_p & f \end{bmatrix}$$

- dim: the dimension of the symbols with the form [w r v s f].

- dontcare the new value of the dontcare symbol (usually -9).

- epsilon the new value of the epsilon symbol (usually -5).

Output argument

- newTable: the table with the new dontcare and epsilon symbols.

waitbarx

Purpose Shows the evolution of a computation.

Syntax
```
wbh = waitbarx('Name bar 1',Value,'Name bar 2',Value,...)
```

Input argument Value is the value of the corresponding bar between 1 and 0 wbh contains the handles of the bar names and their values.

$$\text{wbh} = \begin{array}{c} \text{bar1} \\ \text{bar2} \\ \vdots \end{array} \begin{pmatrix} \overset{\text{Name}}{\text{handle12}} & \overset{\text{Value}}{\text{handle12}} \\ \text{handle21} & \text{handle22} \\ \vdots & \vdots \end{pmatrix}$$

waitbarx(wbh(n,1),'String')
The name of the n-th bar is set to 'String'.

aitbarx(wbh(n,2),Value)
The value of the n-th bar is set to Value .

Example:
```
waitbarx(wbh(3,2),IterationStep/NumberOfIterations);
```

zposs2adj

Purpose Generates the adjacency matrix corresponding to a set of given state sequences.

Syntax `AdjacencyMatrix = zposs2Adj(Zposs, sizeOfZ)`

Input arguments

- Zposs: set of states sequences in a cell-array.

- sizeOfZ: required size of the adjacency matrix.

Output argument

- AdjacencyMatrix: the generated adjacency matrix.

D. Models of the components of the pilot manufacturing cell

D.1. Pusher

z	Description
1	Pusher is pulled back
2	Pusher is extended
3	Pusher is blocked

Table D.1.: States of the pusher

v	Description
1	Pull the pusher back
2	Extend the pusher
0	ε

w	Description
1	Pusher is pulled back
2	Pusher is extended
3	Pusher is blocked

Table D.2.: Input and output signals of the pusher

r_1	Description
1	Pusher is being pulled back
2	Pusher is being extended
0	ε

r_2	Description
1	Pusher is being pulled back
0	ε

Table D.3.: Interconnection output signals of the pusher

s	Description
1	Workpieces on the buffer rail are being shifted
2	Workpieces on the buffer rail are being blocked
0	ε

Table D.4.: Interconnection input signals of the pusher

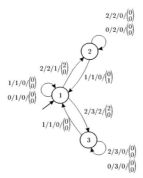

Figure D.1.: Extended Input/Output automaton of the pusher

D.2. Buffer rail

z	Description
1	No workpiece on pick-up position
2	Workpiece on pick-up position

Table D.5.: States of the buffer rail

v	Description
0	ε

w	Description
1	No workpiece on pick-up position
2	Workpiece on pick-up position

Table D.6.: Input and output signals of the buffer rail

r	Description
1	Workpieces on the buffer rail are being shifted
2	Workpieces on the buffer rail are being blocked
0	ε

Table D.7.: Interconnection output signals of the buffer rail

s_1	Description
1	Pusher is being pulled back
2	Pusher is being extended
0	ε

s_2	Description
1	Magazine empty
0	ε

s_3	Description
1	Is workpiece on position ready for the HH gripper?
2	HH gripper picks the workpiece up
3	HH gripper drops the workpiece
0	ε

s_4	Description
1	HH over position 1
$\vdots$	$\vdots$
23	HH over position 23
0	ε

Table D.8.: Interconnection input signals of the buffer rail

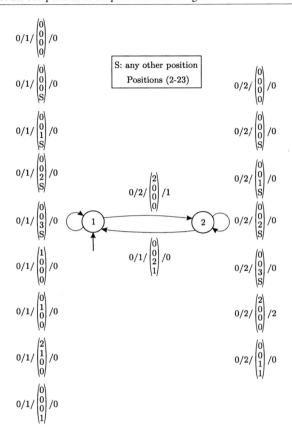

Figure D.2.: Extended Input/Output automaton of the buffer rail

D.3. Magazine

z	Description
1	10 workpieces in magazine (full)
2	9 workpieces in magazine
⋮	⋮
10	1 workpiece in magazine
11	Magazine is empty

Table D.9.: States of the magazines

v	Description
0	ε

w	Description
1	Magazine is full
2	Magazine is empty
3	Magazine is neither full nor empty

Table D.10.: Input and output signals of the magazine

r	Description
1	Magazine is empty
0	ε

Table D.11.: Interconnection output signals of the magazine

s_2	Description
1	Is workpiece on position ready for the VH gripper?
2	VH gripper picks the workpiece up
3	VH gripper drops the workpiece
4	VH does not drop the workpiece
0	ε

s_1	Description
1	Pusher is being pulled back
0	ε

s_3	Description
1	VH over position 1
$\vdots$	$\vdots$
5	VH over position 5
0	ε

Table D.12.: Interconnection input signals of the magazine

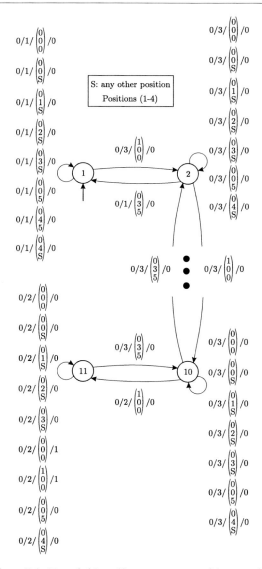

Figure D.3.: Extended Input/Output automaton of the magazine

D.4. Horizontal handler

z	Description
1	HH over position 1
⋮	⋮
23	HH over position 23

Table D.13.: States of the horizontal handler

v	Description
1	Go to position 1
⋮	⋮
23	Go to position 23
0	ε

w	Description
1	HH over position 1
⋮	⋮
23	HH over position 23

Table D.14.: Input and output signals of the horizontal handler

r_1	Description
1	Is HH allowed to move?
0	ε

r_2	Description
1	HH over position 1
⋮	⋮
23	HH over position 23
0	ε

Table D.15.: Interconnection output signals of the horizontal handler

s	Description
1	HH is allowed to move
2	Position of HH is requested
0	ε

Table D.16.: Interconnection input signals of the horizontal handler

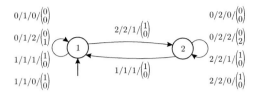

Figure D.4.: Exemplary extended Input/Output automaton of the horizontal handler for two positions

D.5. Horizontal handling gripper

z	Description
1	HH gripper up without the workpiece
2	HH gripper up with the workpiece
3	HH gripper down without the workpiece
4	HH gripper down with the workpiece
5	HH gripper blocked with the workpiece

Table D.17.: States of the horizontal handling gripper

v	Description	w	Description
1	Go up	1	Up without workpiece
2	Go down	2	Up with workpiece
3	Vacuum on	3	Down without workpiece
4	Vacuum off	4	Down with workpiece
0	ε	5	Workpiece can not be dropped

Table D.18.: Input and output signals of the horizontal handling gripper

r_1	Description
1	HH is allowed to move
2	Position of HH is requested
0	ε

r_2	Description
1	Is workpiece on position ready for HH gripper?
2	HH gripper picks the workpiece up
3	HH gripper drops the workpiece
0	ε

Table D.19.: Interconnection output signals of the horizontal handling gripper

s	Description
1	Is HH allowed to move?
0	ε

Table D.20.: Interconnection input signals of the horizontal handling gripper

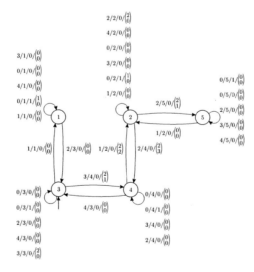

Figure D.5.: Extended Input/Output automaton of the horizontal handling gripper

D.6. Positions in blocks

z	Description
1	No workpiece in position
2	Workpiece in Position

Table D.21.: States of a position

v	Description
0	ε

w	Description
0	ε

Table D.22.: Input and output signals of a position

s_1	Description
1	Is workpiece on position ready for HH gripper?
2	HH gripper picks the workpiece up
3	HH gripper drops the workpiece
0	ε

s_2	Description
1	HH over position 1
$\vdots$	$\vdots$
23	HH over position 23
0	ε

Table D.23.: Interconnection input signals of a position

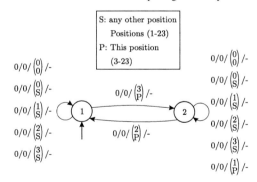

Figure D.6.: Extended Input/Output automaton of a position

D.7. Conveyor belt

z	Description
1	Conveyor belt is empty
2	Workpiece at Sensor 1 (dropping position)
3	Workpiece at Sensor 2 (before slide 1)
4	Workpiece at Sensor 3 (before slide 2)
5	Workpiece at Sensor 4 (before slide 3)
6	Workpiece at Sensor 5 (before slide 6)
7	Workpiece at Sensor 6 (falling from the belt)

Table D.24.: States of the conveyor belt

v	Description
1	Activate the conveyor belt
0	ε

w	Description
1	Conveyor belt is empty
2	Workpiece at Sensor 1
3	Workpiece at Sensor 2
4	Workpiece at Sensor 3
5	Workpiece at Sensor 4
6	Workpiece at Sensor 5
7	Workpiece falling from the belt

Table D.25.: Input and output signals of the conveyor belt

r	Description
1	Deflector 1 is oblique
2	Deflector 2 is oblique
3	Deflector 3 is oblique
4	Deflector 4 is oblique
5	Deflector 1 is straight
6	Deflector 2 is straight
7	Deflector 3 is straight
8	Deflector 4 is straight
0	ε

Table D.26.: Interconnection output signals of the conveyor belt

s_1	Description
1	Is workpiece on position ready for the HH gripper?
2	HH gripper picks the workpiece up
3	HH gripper drops the workpiece
0	ε

s_2	Description
1	HH over position 1
$\vdots$	$\vdots$
23	HH over position 23
0	ε

Table D.27.: Interconnection input signals of the conveyor belt

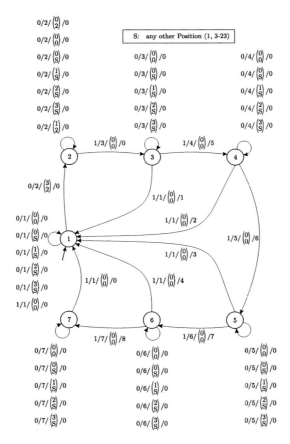

Figure D.7.: Extended Input/Output automaton of the conveyor belt

D.8. Deflectors

z	Description
1	Deflector is straight
2	Deflector is oblique

Table D.28.: States of the deflectors

v	Description
1	Set deflector straight
2	Set deflector oblique
0	ε

w	Description
1	Deflector is straight
2	Deflector is oblique

Table D.29.: Input and output signals of the deflectors

r	Description
1	New workpiece pushed into slide
0	ε

s	Description
1	Deflector 1 is oblique
2	Deflector 2 is oblique
3	Deflector 3 is oblique
4	Deflector 4 is oblique
5	Deflector 1 is straight
6	Deflector 2 is straight
7	Deflector 3 is straight
8	Deflector 4 is straight
0	ε

Table D.30.: Interconnection input and output signals of the deflectors

D: This deflector is oblique

S: This deflector is straight

Q: Other deflectors are oblique

R: Other deflectors are straight

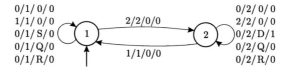

0/1/ 0/ 0
1/1/ 0/ 0
0/1/ S/ 0
0/1/ Q/0
0/1/ R/0

2/2/0/0

1/1/0/0

0/2/ 0/ 0
2/2/ 0/ 0
0/2/ D/ 1
0/2/ Q/0
0/2/ R/0

Figure D.8.: Extended Input/Output automaton of the deflectors

D.9. Slides

z	Description
1	Empty
2	1 workpiece in slide
$\vdots$	$\vdots$
6	5 workpiece in slide
7	6 workpiece in slide (full)

Table D.31.: States of the Slides

v	Description
0	ε

w	Description
1	Slide is full
2	Slide is empty
3	Slide is neither full nor empty
4	New workpiece enters the slide

Table D.32.: Input and output signals of the slides

s_2	Description
1	Is workpiece on position ready for the VH gripper?
2	VH gripper picks the workpiece up
3	VH gripper drops the workpiece
4	VH gripper does not drop the workpiece
0	ε

s_1	Description
1	New workpiece pushed into slide
0	ε

s_3	Description
1	VH over position 1
⋮	⋮
5	VH over position 5
0	ε

Table D.33.: Interconnection input signals of the slides

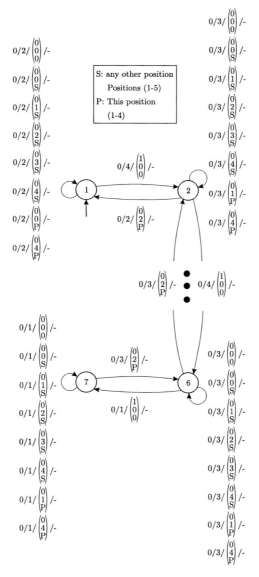

Figure D.9.: Extended Input/Output automaton of the slides

D.10. Vertical handler

z	Description
1	VH over position 1
⋮	⋮
5	VH over position 5

Table D.34.: States of the vertical handler

v	Description
1	Go to position 1
⋮	⋮
5	Go to position 5
0	ε

w	Description
1	VH over position 1
⋮	⋮
5	VH over position 5

Table D.35.: Input and output signals of the vertical handler

r_1	Description
1	VH is allowed to move?
0	ε

r_2	Description
1	VH over position 1
⋮	⋮
5	VH over position 5
0	ε

Table D.36.: Interconnection output signals of the vertical handler

s	Description
1	VH is allowed to move
2	Position of VH is requested
0	ε

Table D.37.: Interconnection input signals of the vertical handler

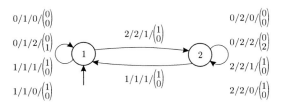

Figure D.10.: Exemplary extended Input/Output automaton of the vertical handler

D.11. Vertical handling gripper

z	Description
1	VH gripper up without workpiece
2	VH gripper up with workpiece
3	VH gripper down without workpiece
4	VH gripper down with workpiece

Table D.38.: States of the vertical handling gripper

v	Description
1	Go up
2	Go down
3	Vacuum on
4	Vacuum off
0	ε

w	Description
1	VH gripper up without workpiece
2	VH gripper up with workpiece
3	VH gripper down without workpiece
4	VH gripper down with workpiece

Table D.39.: Input and output signals of the vertical handling gripper

r_1	Description
1	Is VH allowed to move?
2	Position of VH is requested
0	ε

r_2	Description
1	Is workpiece on position ready for VH gripper?
2	VH gripper picks the workpiece up
3	VH gripper drops the workpiece
4	VH gripper does not drop workpiece
0	ε

Table D.40.: Interconnection output signals of the vertical handling gripper

s	Description
1	is VH allowed to move?
0	ε

Table D.41.: Interconnection input signals of the vertical handling gripper

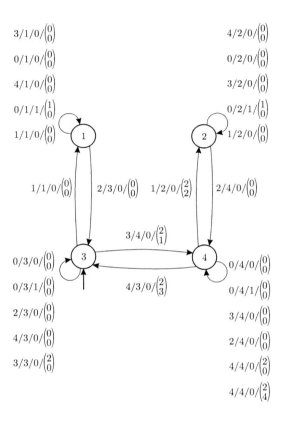

Figure D.11.: Extended Input/Output automaton of the vertical handling gripper

D.12. Encoding tables of all control laws of the pilot manufacturing cell

The following tables describe the meaning of the encoded symbols of the I/O automata network depicted in Fig. 7.6 which models the pilot manufacturing cell of Chapter 7. The symbols in the most left column of each of the following tables represent the encoded states z_c, encoded controller outputs w_c and encoded controller inputs v_c. All symbols used by the controllers $\mathcal{A}_{c-main}$, $\mathcal{A}_{c-SP1}$, $\mathcal{A}^r_{c-SP1}$, $\mathcal{N}_{c-SP1}$, $\mathcal{A}_{c-SP2}$, $\mathcal{A}^r_{c-SP2}$ and $\mathcal{N}_{c-SP2}$ from Chapter 7. Recall that based on (4.3), the following symbols of the controllers are related to those of the plant with $\tilde{z}_p = z_c$, $\tilde{v}_p = w_c$ and $\tilde{w}_p = v_c$. The symbols in columns from P to GVH have to be interpreted with the corresponding tables of states, inputs and outputs defined from Section D.1 to Section D.11.

State encoding table

z_c	P	BR	M	HH	GHH	Pos7	Pos15	B	D1	D3	S1	S3	VH	GVH
1	1	1	1	1	1	1	1	1	1	1	1	1	1	1
2	1	1	1	1	1	1	1	1	1	1	1	1	5	1
3	1	1	1	1	1	1	1	1	1	1	1	1	5	3
4	1	1	1	2	1	1	2	2	1	1	1	1	1	1
5	1	1	1	2	2	1	2	1	1	1	1	1	1	1
6	1	1	1	2	3	1	2	2	1	1	1	1	1	1
7	1	1	1	2	4	1	2	2	1	1	1	1	1	1
8	1	1	1	7	1	1	2	2	1	1	1	1	1	1
9	1	1	1	7	1	2	1	1	1	1	1	1	1	1
10	1	1	1	7	1	2	2	1	1	1	1	1	1	1
11	1	1	1	7	2	1	1	1	1	1	1	1	1	1
12	1	1	1	7	2	1	2	1	1	1	1	1	1	1
13	1	1	1	7	2	2	1	1	1	1	1	1	1	1
14	1	1	1	7	3	2	1	1	1	1	1	1	1	1
15	1	1	1	7	3	2	2	1	1	1	1	1	1	1
16	1	1	1	7	4	2	1	1	1	1	1	1	1	1
17	1	1	1	7	4	2	2	1	1	1	1	1	1	1

. . .

States encoding table

z_c	P	BR	M	HH	GHH	Pos7	Pos15	B	D1	D3	S1	S3	VH	GVH
18	1	1	1	7	5	2	1	1	1	1	1	1	1	1
19	1	1	1	15	1	1	2	1	1	1	1	1	1	1
20	1	1	1	15	1	1	2	2	1	1	1	1	1	1
21	1	1	1	15	1	1	2	3	1	1	1	1	1	1
22	1	1	1	15	1	1	2	4	1	1	1	1	1	1
23	1	1	1	15	1	1	2	5	1	1	1	1	1	1
24	1	1	1	15	1	1	2	6	1	1	1	1	1	1
25	1	1	1	15	1	1	2	7	1	1	1	1	1	1
26	1	1	1	15	1	2	2	1	1	1	1	1	1	1
27	1	1	1	15	2	1	1	1	1	1	1	1	1	1
28	1	1	1	15	2	2	1	1	1	1	1	1	1	1
29	1	1	1	15	3	1	2	1	1	1	1	1	1	1
30	1	1	1	15	3	2	2	1	1	1	1	1	1	1
31	1	1	1	15	4	1	2	1	1	1	1	1	1	1
32	1	1	1	15	4	2	2	1	1	1	1	1	1	1
33	1	1	2	1	1	1	1	1	1	1	1	1	1	2
34	1	1	2	1	1	1	1	1	1	1	1	1	3	2
35	1	1	2	1	1	1	1	1	1	1	1	1	5	2
36	1	1	2	1	1	1	1	1	1	1	1	1	5	4
37	1	1	2	1	1	1	1	1	1	1	1	2	1	1
38	1	1	2	1	1	1	1	1	1	1	1	2	3	1
39	1	1	2	1	1	1	1	1	1	1	1	2	3	3
40	1	1	2	1	1	1	1	1	1	1	1	2	3	4
41	1	1	2	1	1	1	1	1	1	1	2	1	1	1
42	1	1	2	1	1	1	1	1	1	1	2	1	1	3
43	1	1	2	1	1	1	1	1	1	1	2	1	1	4
44	1	1	2	1	1	1	1	1	1	2	1	2	1	1
45	1	1	2	1	1	1	1	1	2	1	2	1	1	1

. . .

States encoding table

z_c	P	BR	M	HH	GHH	Pos7	Pos15	B	D1	D3	S1	S3	VH	GVH
46	1	1	2	1	1	1	1	2	1	1	1	1	1	1
47	1	1	2	1	1	1	1	3	1	1	1	1	1	1
48	1	1	2	1	1	1	1	3	2	1	1	1	1	1
49	1	1	2	1	1	1	1	4	1	1	1	1	1	1
50	1	1	2	1	1	1	1	5	1	1	1	1	1	1
51	1	1	2	1	1	1	1	5	1	2	1	1	1	1
52	1	1	2	1	2	1	1	1	1	1	1	1	1	1
53	1	1	2	2	1	1	1	2	1	1	1	1	1	1
54	1	1	2	2	1	1	2	2	1	1	1	1	1	1
55	1	1	2	2	2	1	1	1	1	1	1	1	1	1
56	1	1	2	2	2	1	2	1	1	1	1	1	1	1
57	1	1	2	2	3	1	1	2	1	1	1	1	1	1
58	1	1	2	2	3	1	2	2	1	1	1	1	1	1
59	1	1	2	2	4	1	1	2	1	1	1	1	1	1
60	1	1	2	2	4	1	2	2	1	1	1	1	1	1
61	1	1	2	7	1	1	2	2	1	1	1	1	1	1
62	1	1	2	7	1	2	1	1	1	1	1	1	1	1
63	1	1	2	7	1	2	2	1	1	1	1	1	1	1
64	1	1	2	7	2	1	1	1	1	1	1	1	1	1
65	1	1	2	7	2	1	2	1	1	1	1	1	1	1
66	1	1	2	7	2	2	1	1	1	1	1	1	1	1
67	1	1	2	7	3	2	1	1	1	1	1	1	1	1
68	1	1	2	7	3	2	2	1	1	1	1	1	1	1
69	1	1	2	7	4	2	1	1	1	1	1	1	1	1
70	1	1	2	7	4	2	2	1	1	1	1	1	1	1
71	1	1	2	7	5	2	1	1	1	1	1	1	1	1
72	1	1	2	15	1	1	2	1	1	1	1	1	1	1
73	1	1	2	15	1	1	2	2	1	1	1	1	1	1

. . .

States encoding table

z_c	P	BR	M	HH	GHH	Pos7	Pos15	B	D1	D3	S1	S3	VH	GVH
74	1	1	2	15	1	1	2	3	1	1	1	1	1	1
75	1	1	2	15	1	1	2	4	1	1	1	1	1	1
76	1	1	2	15	1	1	2	5	1	1	1	1	1	1
77	1	1	2	15	1	1	2	6	1	1	1	1	1	1
78	1	1	2	15	1	1	2	7	1	1	1	1	1	1
79	1	1	2	15	1	2	2	1	1	1	1	1	1	1
80	1	1	2	15	2	1	1	1	1	1	1	1	1	1
81	1	1	2	15	2	2	1	1	1	1	1	1	1	1
82	1	1	2	15	3	1	2	1	1	1	1	1	1	1
83	1	1	2	15	3	2	2	1	1	1	1	1	1	1
84	1	1	2	15	4	1	2	1	1	1	1	1	1	1
85	1	1	2	15	4	2	2	1	1	1	1	1	1	1
86	1	2	2	1	1	1	1	1	1	1	1	1	1	1
87	1	2	2	1	3	1	1	1	1	1	1	1	1	1
88	1	2	2	1	4	1	1	1	1	1	1	1	1	1
89	2	2	1	1	1	1	1	1	1	1	1	1	1	1

Table D.42.: States encoding table of all control laws of the pilot manufacturing cell

Control outputs encoding table

w_c	P	BR	M	HH	GHH	Pos7	Pos15	B	D1	D3	S1	S3	VH	GVH
1	0	0	0	0	0	0	0	0	0	0	0	0	0	0
2	0	0	0	0	0	0	0	0	0	0	0	0	0	1
3	0	0	0	0	0	0	0	0	0	0	0	0	0	2
4	0	0	0	0	0	0	0	0	0	0	0	0	0	3
5	0	0	0	0	0	0	0	0	0	0	0	0	0	4
6	0	0	0	0	0	0	0	0	0	0	0	0	1	0
7	0	0	0	0	0	0	0	0	0	0	0	0	3	0

. . .

Control outputs encoding table

w_c	P	BR	M	HH	GHH	Pos7	Pos15	B	D1	D3	S1	S3	VH	GVH
8	0	0	0	0	0	0	0	0	0	0	0	0	5	0
9	0	0	0	0	0	0	0	0	0	1	0	0	0	0
10	0	0	0	0	0	0	0	0	0	2	0	0	0	0
11	0	0	0	0	0	0	0	0	1	0	0	0	0	0
12	0	0	0	0	0	0	0	0	2	0	0	0	0	0
13	0	0	0	0	0	0	0	1	0	0	0	0	0	0
14	0	0	0	0	1	0	0	0	0	0	0	0	0	0
15	0	0	0	0	2	0	0	0	0	0	0	0	0	0
16	0	0	0	0	3	0	0	0	0	0	0	0	0	0
17	0	0	0	0	4	0	0	0	0	0	0	0	0	0
18	0	0	0	1	0	0	0	0	0	0	0	0	0	0
19	0	0	0	2	0	0	0	0	0	0	0	0	0	0
20	0	0	0	7	0	0	0	0	0	0	0	0	0	0
21	0	0	0	15	0	0	0	0	0	0	0	0	0	0
22	1	0	0	0	0	0	0	0	0	0	0	0	0	0
23	2	0	0	0	0	0	0	0	0	0	0	0	0	0

Table D.43.: Control outputs encoding table of all control laws of the pilot manufacturing cell

Control inputs encoding table

v_c	P	BR	M	HH	GHH	Pos7	Pos15	B	D1	D3	S1	S3	VH	GVH
1	1	1	1	1	1	0	0	1	1	1	2	2	0	0
2	1	1	1	1	1	0	0	1	1	1	2	2	0	1
3	1	1	1	1	1	0	0	1	1	1	2	2	1	1
4	1	1	1	1	1	0	0	1	1	1	2	2	5	1
5	1	1	1	1	1	0	0	1	1	1	2	2	5	3
6	1	1	1	2	1	0	0	2	1	1	2	2	1	1
7	1	1	1	2	2	0	0	1	1	1	2	2	1	1

. . .

Control inputs encoding table

v_c	P	BR	M	HH	GHH	Pos7	Pos15	B	D1	D3	S1	S3	VH	GVH
8	1	1	1	2	3	0	0	2	1	1	2	2	1	1
9	1	1	1	2	4	0	0	2	1	1	2	2	1	1
10	1	1	1	7	1	0	0	1	1	1	2	2	1	1
11	1	1	1	7	1	0	0	2	1	1	2	2	1	1
12	1	1	1	7	2	0	0	1	1	1	2	2	1	1
13	1	1	1	7	3	0	0	1	1	1	2	2	1	1
14	1	1	1	7	4	0	0	1	1	1	2	2	1	1
15	1	1	1	15	1	0	0	1	1	1	2	2	1	1
16	1	1	1	15	1	0	0	2	1	1	2	2	1	1
17	1	1	1	15	1	0	0	3	1	1	2	2	1	1
18	1	1	1	15	1	0	0	4	1	1	2	2	1	1
19	1	1	1	15	1	0	0	5	1	1	2	2	1	1
20	1	1	1	15	1	0	0	6	1	1	2	2	1	1
21	1	1	1	15	1	0	0	7	1	1	2	2	1	1
22	1	1	1	15	2	0	0	1	1	1	2	2	1	1
23	1	1	1	15	3	0	0	1	1	1	2	2	1	1
24	1	1	1	15	4	0	0	1	1	1	2	2	1	1
25	1	1	3	0	1	0	0	1	1	1	2	2	1	1
26	1	1	3	0	1	0	0	2	1	1	2	2	1	1
27	1	1	3	0	2	0	0	1	1	1	2	2	1	1
28	1	1	3	1	1	0	0	1	1	1	2	2	0	0
29	1	1	3	1	1	0	0	1	1	1	2	2	0	2
30	1	1	3	1	1	0	0	1	1	1	2	2	1	2
31	1	1	3	1	1	0	0	1	1	1	2	2	1	4
32	1	1	3	1	1	0	0	1	1	1	2	2	3	2
33	1	1	3	1	1	0	0	1	1	1	2	2	5	2
34	1	1	3	1	1	0	0	1	1	1	2	2	5	3
35	1	1	3	1	1	0	0	1	1	1	2	2	5	4

. . .

Control inputs encoding table

v_c	P	BR	M	HH	GHH	Pos7	Pos15	B	D1	D3	S1	S3	VH	GVH
36	1	1	3	1	1	0	0	1	1	1	2	3	0	0
37	1	1	3	1	1	0	0	1	1	1	2	3	0	1
38	1	1	3	1	1	0	0	1	1	1	2	3	1	1
39	1	1	3	1	1	0	0	1	1	1	2	3	3	1
40	1	1	3	1	1	0	0	1	1	1	2	3	3	3
41	1	1	3	1	1	0	0	1	1	1	2	3	3	4
42	1	1	3	1	1	0	0	1	1	1	2	4	1	1
43	1	1	3	1	1	0	0	1	1	1	3	2	0	0
44	1	1	3	1	1	0	0	1	1	1	3	2	1	1
45	1	1	3	1	1	0	0	1	1	1	3	2	1	3
46	1	1	3	1	1	0	0	1	1	1	3	2	1	4
47	1	1	3	1	1	0	0	1	1	2	2	4	1	1
48	1	1	3	1	1	0	0	1	2	1	4	2	1	1
49	1	1	3	1	1	0	0	2	1	1	2	2	1	1
50	1	1	3	1	1	0	0	3	1	1	2	2	1	1
51	1	1	3	1	1	0	0	3	2	1	2	2	1	1
52	1	1	3	1	1	0	0	4	1	1	2	2	1	1
53	1	1	3	1	1	0	0	5	1	1	2	2	1	1
54	1	1	3	1	1	0	0	5	1	2	2	2	1	1
55	1	1	3	1	2	0	0	1	1	1	2	2	1	1
56	1	1	3	1	5	0	0	1	1	1	2	2	1	1
57	1	1	3	2	1	0	0	2	1	1	2	2	1	1
58	1	1	3	2	2	0	0	1	1	1	2	2	1	1
59	1	1	3	2	3	0	0	2	1	1	2	2	1	1
60	1	1	3	2	4	0	0	2	1	1	2	2	1	1
61	1	1	3	2	5	0	0	1	1	1	2	2	1	1
62	1	1	3	2	5	0	0	2	1	1	2	2	1	1
63	1	1	3	7	1	0	0	1	1	1	2	2	1	1

. . .

Control inputs encoding table

v_c	P	BR	M	HH	GHH	Pos7	Pos15	B	D1	D3	S1	S3	VH	GVH
64	1	1	3	7	1	0	0	2	1	1	2	2	1	1
65	1	1	3	7	2	0	0	1	1	1	2	2	1	1
66	1	1	3	7	3	0	0	1	1	1	2	2	1	1
67	1	1	3	7	4	0	0	1	1	1	2	2	1	1
68	1	1	3	7	5	0	0	1	1	1	2	2	1	1
69	1	1	3	8	2	0	0	1	1	1	2	2	1	1
70	1	1	3	15	1	0	0	1	1	1	2	2	1	1
71	1	1	3	15	1	0	0	2	1	1	2	2	1	1
72	1	1	3	15	1	0	0	3	1	1	2	2	1	1
73	1	1	3	15	1	0	0	4	1	1	2	2	1	1
74	1	1	3	15	1	0	0	5	1	1	2	2	1	1
75	1	1	3	15	1	0	0	6	1	1	2	2	1	1
76	1	1	3	15	1	0	0	7	1	1	2	2	1	1
77	1	1	3	15	2	0	0	1	1	1	2	2	1	1
78	1	1	3	15	3	0	0	1	1	1	2	2	1	1
79	1	1	3	15	4	0	0	1	1	1	2	2	1	1
80	1	1	3	15	5	0	0	1	1	1	2	2	1	1
81	1	2	3	1	1	0	0	1	1	1	2	2	1	1
82	1	2	3	1	3	0	0	1	1	1	2	2	1	1
83	1	2	3	1	4	0	0	1	1	1	2	2	1	1
84	1	2	3	1	5	0	0	1	1	1	2	2	1	1
85	2	2	1	1	1	0	0	1	1	1	2	2	1	1
86	3	1	1	1	1	0	0	1	1	1	2	2	1	1
87	3	2	1	1	1	0	0	1	1	1	2	2	1	1

Table D.44.: Output encoding table of all control laws of the pilot manufacturing cell